AF730529

Copyright © Andrew Romeo & Anthony Sapountzis

The moral rights of the author have been asserted.

All rights reserved. No part of this book may be reproduced, stored in a retrieval system, communicated or transmitted in any form or by any means without prior written permission.

All enquiries should be made to the author.

Get DevReady: A strategy-first playbook to plan, prove, and build software users love

ISBN: 978-0-646-85016-0

Disclaimer

The material in this publication is of the nature of general comment only, and does not represent professional advice. It is not intended to provide specific guidance for particular circumstances and it should not be relied on as the basis for any decision to take action or not take action on any matter which it covers.

Readers should obtain professional advice where appropriate, before making any such decision.

To the maximum extent permitted by law, the author and publisher disclaim all responsibility and liability to any person, arising directly or indirectly from any person taking or not taking action based on the information in this publication.

Get DevReady

A strategy-first playbook to plan, prove, and build software users love

Andrew Romeo & Anthony Sapountzis

Table of Contents

Foreword

When I started MyHome in the UK back in 2000, I had one goal: to build a branded, consistent, scalable premium service business that was a fully templated, systemically driven, turnkey operation that could be rolled out. We were so successful, that within 5 years we had over 900 outlets in 9 European countries and went on to successfully float the company on the London Stock Exchange in 2006. Some years later, I brought the model to Australia, where I focused on refining the system—taking advantage of all the latest technology to remove as many friction points, to make the business as simple to run as possible for our partners.

Myhome offers its franchise partners a proven pathway to build successful businesses and still have time for the important things in their lives - spend time with their families and doing things they want and like to do. True work life balance.

At MyHome, we have a saying that 'Success is in the System' that the key to truly have a business that works for you rather than the other way round was taking a systemic approach. We have always been a fanatical about systems and this has been the key to our success but as technology became more readily available and affordable I saw an opportunity. I could see that driving our processes and systems digitally would take the business to a totally new level. We had a vision for having an operating system that would give our franchise partners and the support team here at head office everything they needed to easily and intuitively make the business run Smoothly. All is one place.

We searched the world for software we could take off the shelf and use, but nothing matched what we knew we needed. So, we determined that to make the investment and create something from scratch to meet our immediate needs and grow with the business as we scale.

It was a big move so finding the right IT partner was crucial. We did our research and conducted an intensive beauty parade of potential suppliers and eventually made the decision to partner with Aerion and we have never looked back.

Back then, we just had a single location in Hawthorn. Today, we've grown to 20 locations across all states — and counting. That growth has been made possible by the technology backbone we built together.

Aerion don't just write code, they are our partners in all aspects of IT. What impressed me most was that they took the time out - over many long meetings, to unpack our processes, align with our vision and really understand our business. Together we mapped out the whole program, expectations were clearly defined, timelines and costings were all set - all before the first line of code was written. We ended up with a blueprint so I could see right up front what the end result was going to be and what it was going to cost. This gave me the comfort and confidence we needed and make the investment to start the journey.

That's the power of the Aerion DevReady Process.

Today, we have a platform that truly supports MyHome's end-to-end operations—from lead capture and quoting, to team scheduling, payment processing, and even AI-powered quality control. The platform isn't just "tech"—it's a fully-fledged business engine. Ready to scale.

This book outlines that process in full. And if you're a founder or operator who's ever felt lost in the world of software development—this is your guide. DevReady strips away the noise. It helps you clarify your idea, understand your market, avoid costly mistakes, and most importantly, build something that actually works. The journey of building tech is never easy. But with the right people, the right process, and a mindset grounded in clarity and execution, it is absolutely doable.

MyHome is proof of that and with the help of Aerion and DevReady you too can turn your vision into reality.

Russell O'Connell

Founder & CEO, MyHome

CHAPTER

1

Getting DevReady (The Right Way to Launch)

The year was 2016 and the idea was named FinalCall.

A technological platform developed by a business analyst and project manager named Paul Arnost, FinalCall had the potential to revolutionise real estate transactions. The platform would enable digital transactions of high-value assets, such as property, while also enabling digital negotiations. And like all great ideas, it came about when Paul saw a problem in the market that nobody else seemed interested in solving.

He'd watched several of his friends go through the house-hunting process, with each of them becoming more and more frustrated with the slow progress made with their transactions. Whether it's rentals, selling, or buying, working with real estate has always been a laborious and intensive process that requires an enormous amount of manual work. There isn't a lot of transparency in the industry and buyers, in particular, can spend months, if not years, of their lives going to house viewings and trying to set up their deals. And when they finally make a bid, based on all of the information they've received from the agents they've worked with, it gets knocked back without a second thought and the buyer has to start all over again.

It's such a painful and time-consuming experience.

Add in the more recent issues related to the coronavirus and how it has limited real estate transactions and you have an industry that's seemingly primed for a digital solution.

Paul looked at the existing process and realised that so much of it involved dead time where nobody does anything. Buyers get frustrated, they waste time, money, and ultimately, the home-buying experience isn't the joyful time that it's meant to be.

The seed of FinalCall was planted.

What if there was a platform that stripped away all of the dead time and made it easier than ever for buyers to negotiate deals with sellers? What if people could conduct the majority of the transaction online using a technological solution, rather than spending time, money, and effort going to house viewings and attending auctions?

Excited by these new ideas, Paul was ready to move forward.

Unfortunately, he had a few hiccups along the way.

The Two Hiccups That Held Final Call Back

Market research is always the first step a budding entrepreneur engages in when launching a new product.

However, it's a step that Paul Arnost skipped almost entirely when developing FinalCall. His excitement over the idea of revolutionising the real estate industry got the better of him and he launched straight into the design and development of the product. Looking back, Paul says that he was probably a little overzealous with the idea. But at the time, he felt like he was on to something and wanted to get it to market as quickly as possible.

Perhaps the biggest unforeseen issue that Paul encountered was that he approached the product from a consumer's perspective. In doing so, he underestimated the need to get estate agents on board for the idea. Paul operated under the assumption that agents would love FinalCall because they could potentially make more money out of using it. They would see it as a tool they could use to maximise their earning potential.

while developing FinalCall. If he were to do it all over again, he says he would solicit far more feedback from real estate agents to discover if FinalCall was something they'd even want to use.

Beyond industry research, Paul says that finding the right people for the job is also a critical component of developing a technological solution. In his case, he selected a team that seemed capable of producing good results. But the experience of working with them felt like pulling teeth from day one. Every issue that arose seemed to result in a shortcut workaround from the developer rather than a genuine solution to the issue. And the defensiveness he faced when he tried to confront his team about the issues only added to his frustration.

After learning his lessons the hard way, Paul decided to essentially start all over again with FinalCall. He decided to leave his job and focus on the product full-time. He went back to the drawing board on his research, soliciting feedback from as many agents and real estate company heads as he could so he could tweak FinalCall to their liking. And eventually, he managed to develop a version of FinalCall that he could put out to the market.

Still, Paul's story is a cautionary tale for many entrepreneurs.

It shows us that having a great idea does not necessarily mean that you will develop a successful product. A lack of research, combined with the wrong team, can result in enormous amounts of wasted time and money.

These are the sorts of challenges that so many non-technical entrepreneurs face. And they're the challenges we aim to help you to overcome with the DevReady process.

But we'll get to that in good time.

First, we want to share another story. Only this time, we want to show you what developing a technological solution looks like when it's done the right way.

Will Strange and Sports Performance Tracking

Will Strange is a pretty competitive person. He grew up playing a bunch of different sports and he knew that he'd need to find something to whet his competitive appetite when he left school. Luckily, he also happened to be a great talker. Granted, not everybody wanted to listen to him! But he had the gift of the gab combined with a competitive edge.

Real estate seemed like a perfect fit!

As a pimply-faced 18-year-old, he found himself selling commercial and industrial property. Over time, he transitioned into property development and had a career that was going pretty well.

But he wanted more.

When he reached his mid-20s, Will realised that he needed to push harder if he was ever going to make something of himself. He needed to take the lessons he'd learned from working in real estate and apply them to his business ideas. This revelation marked the birth of Three65 Underwear, which was an underwear subscription business aimed at men in sports. The idea came to him when he'd just moved away from home and found out just how expensive underwear can be. The subscription model seemed to be the perfect solution, both because it would provide people with affordable underwear and because it overcame the issue of people simply forgetting to buy new underwear.

Ultimately, Will says he developed a product for himself and introduced it to a larger market.

Much to his surprise, he found that most of his customers were women, even though he'd set the business up to focus on men. And this was just the first of many lessons he learned while running his own company. Thanks to Three65 Underwear, Will got a crash course in everything from building websites to satisfying different stakeholders.

Then came the appearance on Shark Tank.

Two years after founding his business, Will found himself on a TV set, pitching his company to the sharks. He saw it as an opportunity to expose his company to an audience of about 700,000 people. Perhaps less expected, he ended up walking away from the show having made a deal for investment from both Janine Allis and Naomi Simson. In the days that followed, he went through the due diligence process with the investors successfully.

But there was a challenge ahead of him.

In addition to Three65, Will had also created a business called Sports Performance Tracking. And at the time of his appearance on Shark Tank, he was closing a capital round on this new business that was much closer to the seven-figure mark than the investment he'd received into Three65.

Now, he faced a choice.

Divide his attention between both businesses, which created the risk that neither would get everything they needed from him. Or, sell one of the companies so he could give his undivided attention to the other.

Will chose the latter option and made the difficult decision that he had to leave Three65. Ultimately, he decided not to accept the Shark Tank duo's offer of investment and he started looking into passing Three65 on. Initially, this involved working with his current investors to find somebody who could run the business in his stead. But that proved to be a nightmare as he just couldn't find the right person for the job. A meeting with a business broker followed soon after and, before he knew it, Will was in the process of selling the business that had taken him from his garage to a national TV audience.

It was a difficult time for Will.

Here he was, selling the first business he'd ever started. He'd built the entire thing from scratch and, in some respects, it felt like he was giving up a limb on his body so he could focus on something else.

Still, he knew that Sports Performance Tracking held much more potential than Three65 Underwear. Will went through with the sale and finally started focusing his full attention on his new idea.

What is Sports Performance Tracking?

What we haven't mentioned yet in this story is that Will never gave up on the sports that he loved so much. While running Three65, he was also playing Australian football, for a local team. Will's quick to point out that he was far from the best player on the team. But it was during this time that he started dabbling with the array of fitness gadgets that have entered the market in recent years, such as the Fitbit and some of Garmin's pieces of tech. But he found that none of those gadgets gave him what he needed as a footy player.

And so, an idea was planted.

What if he could combine his love for his sport with a technological business idea that helped the players?

From this seed, Sports Performance Tracking sprouted.

Much like with Three65, the initial concept for the idea revolved around Will building a product for himself. He wanted something that would track how many kilometres he ran, and at what speeds, during training. What's more, he wanted the product to track if he was doing enough in training to prepare himself for his games. Essentially, he needed something that would act as a fitness GPS.

But Will didn't immediately go out and start developing that product.

Instead, he started looking for products that already existed on the market. That's how he came across a wristwatch that did a lot of the things that he wanted his new product to do. Will bought the watch, learned, and decided that he wanted to get involved with it. Working with the original supplier, Will made some changes to the watch. He pulled the face off it and repackaged it with his own branding.

Eventually, he became the biggest reseller for that product in the world.

And this is the biggest take away from this part of Will's story.

He could have developed that watch himself. He could have taken his idea, sunk millions of dollars into developing a product, and assumed a ton of risk in his fledgling business. But he didn't. Rather, Will found an existing product and built his business based on what that product taught him.

The watch that Will found served as his minimum viable product. His experience as a reseller allowed him to see that there was a big market for the technology that the watch had to offer. But he also knew that the watch did the bare minimum to serve that market. Will's next step was to iterate on that original idea so he could build a product that served the market even better.

Will started working with the technology that he'd discovered, developing new products based on the watch he'd been so successful reselling. Each of these prototypes went through testing periods, with scores of players trying them out and offering feedback. This acted as Will's crucial product research, which is an area that Paul Arnost identified as one that he failed to recognise the importance of. These trials led to the development of what Will affectionately calls a "man bra", though it's probably better defined as a vest that athletes can wear while training.

Will's initial trials produced some interesting results. He could see straight away that many of the players who used his devices started competing with each other. If one player ran 14km in one training session, others might try to top that in later sessions. So, the device definitely acted as a way to push athletes to train harder.

The challenge was that this desire to compete against one another only extended so far.

Once the athletes saw they were getting consistent results from their training, they stopped caring about the product as much. If somebody knows they're going to run about 14km in training every time, they do not need the tracker. Will went back to the drawing board to figure out how he could make his tracker more interesting than a simple GPS device that fed distances back to players.

Ideas started coming thick and fast. Finally, Will went with the idea of making the device track acceleration and deceleration, as doing so would help athletes track their fatigue levels. Somebody who accelerates and decelerates a couple dozen times in a 400m run will feel more tired than somebody who runs 5km smoothly. This general concept allowed Will to come up with the idea that looking into the data behind *how* people train was as important as just measuring running distances and other basic stats.

It was also at this point that Will realised that he couldn't build a sustainable product by relying on the tech that he bought from other suppliers. By selling somebody else's product first, he'd established his minimum viable product. He'd also done it while assuming little to no risk himself. It wasn't Will pouring money into the development of the product. It was the company that he acted as a reseller for. They'd invested in the tech, which allowed Will to focus on coming up with products that could use that tech in better ways.

But with the minimum viable product established, Will needed to change his strategy.

To build Sports Performance Tracking into what it had the potential to be, he would need to start developing his own tech. And for that, he needed the help of investors.

And this brings us to the point where Will was juggling Three65, a proven business concept, with Sports Performance Tracking, a concept that had enormous potential.

Building a Business

With that realisation, Will started building a business around his new idea. Along the way, he recognised that hardware comes with its own issues. Some of his devices wouldn't work in certain temperatures, which required him to redesign based on specific needs. Will underwent an intensive six-month testing process, to ensure he had a device that could work consistently, regardless of who used it or where it was used.

Google became Will's best friend as he devoted hours of his time to research. He also started leaning on his network, strengthening connections with people who could help in the design and sale of the product. And most importantly, he found a software developer who could deliver what he needed from the product.

Ultimately, months of hard work and research led to the development of a proven concept that was ready to bring to the market. And because Will had proven his idea was effective with his minimum viable product (MVP), he was able to attract greater investment and feel more confident in the success of his technology upon launch.

Like many business owners, Will's progress slowed down in the wake of the pandemic. But he's optimistic about the future. Sports Performance Tracking enjoyed a successful launch and it looks set to grow enormously in the coming years.

Today, Sports Performance Tracking has over 30,000 customers, which include major sports teams, clubs, and leagues. The company sells its products in over 100 countries, with each individual country amounting to hundreds, or even thousands, of sales. And Will did it all by using somebody else's watch to establish a minimum viable product before building on what he'd learned to create a successful business.

The key here is that you note the contrasts between Paul's and Will's stories. With Paul, we have an example of an entrepreneur who admits that he jumped straight into design and development without doing as much research as he needed to. Paul also ended up choosing the wrong people for his team, which set him back even further. But with Will, we have an example of somebody who took his time, did his research, and ended up with a much smoother launch as a result.

Will got himself DevReady before bringing his product to market. And that is the critical difference between Paul and Will.

We're delighted to say that Paul has managed to overcome the initial hurdles he faced and now has a successful product of his own. But his story serves as a cautionary tale for anyone who wants to develop and release a tech-based product. If you don't follow the right process, you can create more challenges than you need to face. And the result, in all cases, is the loss of time and money as you attempt to bring your idea to market.

We created the DevReady model to help people like you sidestep the challenges that Paul Arnost faced with FinalCall. And soon, we're going to introduce that model to you and talk about how you can leverage it to bring your own ideas to life. But before we do, you need to know a little more about who we are and what we do.

CHAPTER

2

Who Are We (And Why Do We Want to Help You Get DevReady)?

Waste! Waste! Waste!

When it comes to software, be it developing or using software solutions, waste is the word that we believe should be on everybody's lips. The inability to identify what software you need, who can develop appropriate solutions, and how to leverage the software you have leads to enormous amounts of wasted time and money.

And it's not just us who believe this.

An article published on CIO.com in 2016 highlighted the results of a study that found the average American company deals with a shocking amount of software waste every year. The study, conducted by real-time management platform providers 1E, examined 129 companies from 14 industries. Researchers looked at 3.6 million desktops and 1,800 software titles installed across them over a four-year period.

The goal.

To determine if the average business leverages its software to its fullest.

Unfortunately, the answer is that the vast majority don't. Upon concluding the study, 1E found that the average desktop computer generates software-based waste of $259 per year. Multiply that by 3.6 million desktops and you have over $9 billion of waste money every year! And what's more, the study found that this waste didn't change much over five years. Every industry demonstrated software-based wastefulness at some level, with even the most efficient still wasting 8% of the software it has installed on desktops.

How does this end up happening?

At the basic level, it comes down to businesses buying software that they either don't need or won't use to the fullest. Digging deeper, we can identify internal conflicts as a big issue. A business unit may identify software that does everything they want only to have IT disagree with implementing said software because it doesn't work how they want it to work. The result is almost always a compromise that leads to the creation of software waste.

As an entrepreneur or business owner, this struggle to use software efficiently creates an enormous cost burden for your team. But this isn't where the software waste conundrum ends. Right now, you have an idea for a piece of tech that you want to build a business around. Developing this idea will require you to create a team and confront the software issues we've just discussed. However, you must also face another area of software waste that 1E's study doesn't talk about.

Wasting time and money on a piece of software only to find that it doesn't do what you need it to do.

Think back to Paul Arnost's story from the first chapter. Paul had a phenomenal business idea from a consumer standpoint. His FinalCall platform smooths out so much of the friction that both buyers and sellers have to deal with in the traditional real estate transaction model. It made negotiating easier than ever before and provided buyers and sellers with a means to communicate instantly rather than having to wait for messages to get passed around by real estate agents. The product even has features that make it ideal in today's world, especially when we think about how much damage the coronavirus pandemic has caused to the real estate industry.

And yet, the first iteration of the FinalCall failed.

In jumping straight to development, Paul Arnost didn't consider the needs of the people who would have the most significant influence on whether the product succeeded. Combine that with Paul choosing the wrong people to develop the first version of the software and you have $190,000 down the drain and 10 months spent developing a product that eventually wound up going back to the drawing board.

Thankfully, Paul was able to take advantage of some grant programs and bring his idea to fruition. But the sad truth is that most entrepreneurs don't have the resources needed to absorb such a huge amount of waste.

That's where Aerion Technologies comes in.

We founded this company to confront these issues of software waste on all fronts. And now, we'd like to tell you a little more about us.

Who Are We?

Clarity, growth, inspiration, contribution, and collaboration.

Upon founding Aerion Technologies, we chose these values as the basis of everything that we do. Our names are Andrew Romeo and Anthony Sapountzis and our goal is to help people like you bring their digital solutions to the market to ensure the success of your product.

Aerion Technologies is a software consultancy and development firm. Our role is to help you take your ideas and convert them into software solutions that provide efficient gains, avoid the issues of waste we talked about earlier in this chapter, and, ultimately, help you to create a product that gives value to your customers.

We help non-techs build great tech through custom software development and the DevReady framework we're about to introduce.

Take the example of Shopfront Solutions, a SaaS platform that we developed to serve over 1,500 pharmacies.

Shopfront Solutions exists because we decided that we needed to better understand the complexities and challenges non-tech entrepreneurs face when commercialising their solutions. This resulted in us learning about end users, scaling, stakeholder management, product sales, partners, distribution channels, and much more.

In 2015, we started working with a client named Symbion.

A national healthcare services and products wholesaler, Symbion had over 4,000 pharmacy customers in Australia. For context, there are about 5,700 pharmacies in Australia, indicating that this was a big business.

Symbion managed the marketing for several independent stores, much of which was done manually, when we started working with them. The company's key challenge lay in educating pharmacy store managers about how to use technology to create consistent marketing materials that were high-quality and on-brand.

Partnering with Symbion, we built the first version of a product that would solve this challenge. With a minimum viable product established, we then worked directly with end users to gain feedback that we could use to improve the fledgling product. After several years of going through this process, Symbion decided to commercialise their marketing platform.

That's when Shopfront Solutions was born.

Today, Shopfront Solutions leads the way in retail marketing. It provides a platform designed for retail stores that allow them to manage their print and digital marketing. Those using the platform gain access to a dedicated support team, a ticketing system that allows for easy raising of issues, and a host of marketing materials and research.

And we're not done yet.

Together with Symbion, we're continuously improving the model we've created for Shopfront Solutions so that it adds greater value and in store automation for users. Shopfront has even launched the Pharmacy View Podcast, which it uses to provide value to its clients beyond the Shopfront Solutions platform.

This is just one example of the many projects we've worked on.

We've had the privilege of working on a wide array of software development projects, from multi-billion global initiatives to start-up concepts where challenges related to lack of capital and the need to break into a market are at the fore. In all cases, we've seen the profound impact that having an effective software solution can have in terms of bringing a vision to life and executing it at scale.

Ultimately, we want to help people like you take your ideas from conception to successful launch while facing as few roadblocks as possible along the way. And we've developed a methodology that will help you to do just that. We call it DevReady and it's our way of exploring every possible scenario within a project to deliver a clear development and launch plan.

DevReady is the methodology that we've used to reduce software waste and enable successful launches for so many of our clients. And in this book, we're going to unveil the DevReady methodology secrets and show you why this process will help you take your digital solution from concept to reality.

It's time to get DevReady.

CHAPTER

3

Introducing the DevReady Process

An international engineering firm that operates across two continents, Aurecon is a massive player in its space. The company generates over $1 billion in revenue every year and maintains a vast team of 7,500 employees.

But in 2015, Aurecon faced a challenge.

The company committed to making a big push towards improving its digital offering so it could enhance its client experience. Aurecon needed to digitise its infrastructure to make that happen.

So, they partnered with us.

Working together with the Aurecon team, we developed five solutions that were deployed throughout the business. These solutions included an ideation platform, an asset management system, and an onsite inspection tool. We also worked with Aurecon to develop a site selection tool focused on the telecommunications industry.

However, that isn't the end of the Aurecon story.

Building from our earlier successes, we now work with Aurecon's digital futures team to create a range of products that will enhance their offering.

The first of these digital products is Aurecon's Field Force App.

So…

What does Field Force do?

Rail network operators around the world all face the same challenge of creating, maintaining, and optimising highly complex rail networks. Typically, this has always been a manual process, requiring a great deal of work hours to collect inspection data on critical rail assets, such as switches, tracks, signals, stations, overhead wiring, and facilities.

The Field Force app streamlines this costly and time-consuming process.

The platform allows inspectors to quickly capture all of the relevant information about a rail network asset while on-site. The collected data gets sent to a management interface, which helps other users to generate custom reports in a range of formats. In the space of a couple of clicks, network operators can access all of the data they need to make crucial decisions about their rail networks.

The Field Force roll out has been a complete success.

The app is currently used in 10 countries and has 2800 individual users. To date, it has collected 192,520 images and over 1,700,000 data values. Rail network operators have used this data to create 1,939 forms and generate 15,463 surveys.

This was a wide-reaching technology project that had a lot of moving parts. Of course, many moving parts mean many opportunities for things to go wrong. But with the DevReady process, combined with Aerion's ongoing support, Aurecon now has an app that streamlines its processes, empowers employees, and has enabled them to win new projects.

Why We Designed DevReady

Aurecon's Field Force App showcases the DevReady process in action. With our help, the firm took a concept from the ideation stage through to delivery.

That's what DevReady is designed to do.

But why do we do this?

That's a question that we've faced countless times, especially since the publication of Simon Sinek's landmark book *Start With Why*. And to answer it, we want to take you on a bit of a journey.

The year is 2008 and four guys named Andrew, Anthony, Gordon and Stephen decided to found a business called Aerion. We were fresh out of university, with none of us having any sort of experience in the business world. We were just four young and naïve guys who believed we could create something that made a difference.

Our first project came when we partnered with VSSEC (Victorian Space Science Education Centre). The challenge was to build a simulation game that students in Years 9 and 10 could play. The goal of the game was to keep a group of astronauts alive on the surface of Mars, which was actually a mock surface based in VSSEC's Strathmore centre.

The project was a lot of fun.

It also served as a crash course in our own business. Along the way, we made many mistakes, went over budget, and struggled to pay ourselves. We also had to part ways with a good friend and co-founder of the business.

But we stuck at it and delivered a highly successful outcome. This project, which wasnamed "Mission to Mars", is still in operation and has educated over 50,000 students to date. You'll also be learning a lot more about it later in the book.

It remains one of our proudest achievements.

In the years that followed, we worked on a wide range of solutions. Some hit the mark, delivering real value for our clients. Others, sadly, became shelfware, even though they were good pieces of software. That was heartbreaking to us. In every project, we focused on following our clients' lead. Perhaps that was a little of our business naivety coming into play, as clients occasionally took the solutions we created down paths that we didn't believe were the right paths to follow. It's so disheartening to pour your heart and soul into writing hundreds of thousands of lines of code, only for the resulting software to get shelved, never to see the light of day.

We felt frustrated.

But through that frustration, a crucial question arose:

Why?

Why do projects fail?

That question became our obsession.

We spent months reviewing all of our projects to figure out why the best projects went well and why the failed projects fell apart. Finally, we had our light bulb moment. The reason why so many projects fail is a lack of clarity.

Clarity is the key to success in any digital enterprise. And that revelation led to the creation of DevReady.

Introducing the DevReady (And Delivery) Process

If clarity is the solution to our project woes, we now face another question.

How do we create clarity for our clients so that the solutions that we work so hard on developing actually see the light of day?

The answer to that question lay in the creation of a process.

DevReady is a co-creation process that brings key stakeholders to the table to ensure everybody involved in a project understands its key challenges, needs, and possibilities. It's an 11-stage process, which we split into three core areas:

1. DevReady

2. Building a Team

3. Delivery

Each of these areas has stages to follow that enable a business to take its digital concept from ideation to commercialisation. We call it the DevReady Process and it looks like this:

Area #1 - DevReady

Five stages fit into the DevReady area of the process:

Stage #1 - Ideate

Before you can even think about developing a software solution, you need to gain clarity on the core idea that underpins said solution. That's where the Idea test age comes in. Here, you will devise the Big Idea that you believe will shake up your niche and offer massive value to your clients. You also evaluate and analyse the Big Idea, looking for any challenges you might face while checking to ensure the Big Idea is going to be as impactful as you hope it will be.

Stage #2 - Investigate

You've got a Big Idea. You feel comfortable in the fact that the idea will have an impact on your market. Now, you undergo a period of investigation to confirm those feelings. This stage is all about conducting market research, which includes learning about similar products and potential competitors in your chosen market.

Stage #3 - Design

With your investigation complete, you feel like you're ready to start developing your solution. But before you can get into the practical aspects of the work, you've got to design it. This is all about turning the Big Idea into something tangible.

We break this stage into two parts:

1. Documenting the project, which includes outlining the vital processes and features to come up with a minimum viable product (MVP).

2. Visualising the design so you get an idea of how it will work in practice. You'll consider the user flow, how the interface will work, and build wireframes here.

Stage #4 – Estimate

How much is all of this going to cost?

That's the question we answer at this stage.

One of the reasons that so many digital projects fail is that cost estimates can vary wildly. In many cases, organisations fail to take critical costs into account, which leads to projects running over budget. Here, we outline precisely what you need to consider when creating your estimate, ensuring that you have a realistic picture of how much your project will cost and whether you're capable of covering those costs at this moment in time.

Stage #5 – Validate

All of your research suggests that the Big Idea is going to have a significant impact on your market. You're also confident that you can cover the costs and you have all of your design documentation laid out.

Now, it's time to validate the Big Idea.

This involves getting validation for the timing of the idea and the business case behind it. This stage also involves getting your project's primary stakeholders on board and conducting user testing.

Area #2 – Building a Team

This stage is so vital that it gets a whole section dedicated to itself.

The sixth stage of our process focuses on what you need to do to build a team that can bring your solution to a broader audience. It covers how to get the right tech people on board, what qualities you need to look for in the people who will develop and deliver your solution, and how to work with your tech team to ensure the project comes to fruition.

By getting your team on your side and aligned with the project, you'll experience a much smoother delivery process.

Speaking of which…

Area #3 – Delivery

With the project designed and your team built, it's time to get it on shelves so it can deliver the impact you believe it can have. There are five more stages here, which build on the six you've already worked through:

Stage #7 – Engineer

Great product engineers can make or break the success of a digital endeavour. This stage focuses on how to find a product engineer that can help to bring your solutionto fruition. We cover the qualities to look for in your engineers and how to make sure you have the right person (or people) on board.

Stage #8 – Develop

Up to this point, you've put a lot of time and effort into research, design, and building the right team for your project.

Now, it's time to develop it.

Here, we focus on creating a project timeline, writing the code, and building your testing procedures. You'll start working towards making your MVP and start paying attention to the scalability of the project over time.

Stage #9 – Quality Assurance

In the previous stage, you developed your testing procedures. Here, you put them into practice.

Failure to test a solution properly before launch can lead to the delivery of something that breaks when your clients try to use it. This destroys the trust they have in your product, which can lead to the project faltering before it ever gets off the ground. This stage examines how to build a quality assurance team and what they need to do to test your product correctly so that it delivers on both your and your clients' expectations.

Stage #10 – Product Adoption

Having a great product doesn't mean anything if people aren't using it.

At this stage, we examine the critical reasons why projects fail and help you to raise your adoption rate. This is all about building a user base that makes your solution commercially viable.

And that brings us to the final stage of the process.

Stage #11 – Commercialisation

The product is up and running. You've designed and developed something that serves a need and has the potential to make a real impact on your market. Now, it's time to commercialise it so that your solution reaches its full potential.

This final stage helps you to build a marketing plan, create a pitch deck for use when seeking investment, forming strategic alliances, and building your pricing model.

How Does DevReady Differ From the Typical Development Process?

So, we see that our process contains eleven stages that take you from coming up with the Big Idea to turning that concept into something that you can commercialise. As the book progresses, we'll work through each of these eleven stages, providing you with everything you need to know to execute them effectively.

But there's one more question we need to answer in this chapter.

What makes DevReady so different from the many other project development processes out there?

We've designed the DevReady process to be a non-intimidating form of work shopping that allows you to nail down exactly what is needed to create a commercially viable digital product. The process begins with a general discussion about the wants and needs related to the project, which helps you to develop all of the documentation that sits behind the intended product.

The key with DevReady is that it prompts you to open up the discussion so you take it further than you might with other project development processes. We encourage you to break down every single manual process that you need to automate. DevReady helps you to understand the different types of people who need to use the system and why they need to use it. Think of DevReady as a journey that empowers you to view the problem your product solves through an array of lenses, which allows you to develop a comprehensive and scalable solution.

Think of DevReady as being similar to typical business analysis.

The difference is that it's done using much less formal methods. This means the intimidating aspect of a business analysis, which often feels very formal and rigid, gets replaced by a free-flowing structure that allows for idea sharing and innovative thinking. DevReady is designed to instigate conversations that help you to clarify more than just how you're going to bring a solution to the market. It also allows you to focus on why that solution is needed in the first place.

Conversations do not start with the solution.

They start with a thorough understanding of what the problem is and why your market needs to solve it.

DevReady gives you clarity on these issues before providing you with a straight forward process to follow so that you can create a product that offers the solution your market is actually looking for.

Let's Get Dev Ready

DevReady is all about clarity.

Clarity in terms of why your market needs the solution that will grow from your Big Idea.

Clarity in the way that you're going to design the solution so that it fulfils all of the needs you identify.

And clarity when it comesto building a team, developing the solution, and commercialising the resulting product so that it solves the problem for the widest audience possible.

In creating DevReady, we focused on a "why" question. We wanted to figure out why so many projects, despite resulting in the creation of outstanding solutions, ended up failing. Our investigations of our previous work helped us to come up with the answers. And in the DevReady framework, we've created something that allows others to turn their concepts into commercialised digital solutions that the market needs.

In the coming chapters, we're going to break down every stage of our process. By the end, you'll know exactly what you need to do to create a DevReady solution that's right for your market.

Don't rush into development before you're DevReady.

Keep reading and learn more about the framework that empowers you to create digital solutions that stick.

CHAPTER

4

Stage One – Ideate

You may remember VSSEC from the previous chapter. The company aims to inspire students in Victoria, Australia, to embrace scientific learning and literacy. Its focus lies in STEM (Science, Technology, Engineering, and Mathematics) education, with its mission being to lead students towards an enriching and fulfilling career in the sciences.

For years, VSSEC maintained a Mars educational program that empowered its students to explore worlds beyond our own. As you'll recall from earlier, they eventually decided they wanted to ramp that program up.

The challenge was to develop a simulated Mars Rover mission.

VSSEC wanted to develop a miniature model of the Mars Rover, which its students would then be able to control during simulated scenario-based robotic missions.

Why?

VSSEC's mission is to inspire students with the sciences. This project would enable it to utilise robotics and software to create an engaging platform in which students could work cooperatively to achieve predefined mission objectives. So, in addition to providing students with a deeper understanding of the technology behind the Mars Rover, the project would also give them some experience of what it may be like to work in a science-based environment as part of a team.

VSSEC engaged Aerion to deliver the simulation technology to support its Mars Rover robot.

The challenge was to deliver a software package that simulated a "Mission to Mars" involving two groups of students:

1. Astronauts

2. Mission Control

One group of students would dress up in space suits and walk around the simulated Mars surface to work. This surface was 20 metres wide and 10 metres long, with sand rocks that participants can feel under their feet. There is even a dome roof that uses special red lights to simulate the atmosphere on Mars.

Mission Control relayed instructions to those students. Each member of the Mission Control team has a specific role that the software needs to account for.

Adding to this was the major challenge of the project. We needed to figure out how the program we developed for VSSEC would interface with their Mars Rover, particularly in terms of driving and tracking. This involved understanding what the Rover was designed to do, allowing for the creation of software that relayed instructions within the robot's capabilities.

With VSSEC, we followed the DevReady process.

The Big Idea was already established, which allowed us to move straight into working together on the design for the solution. When we moved into the development portion, we found that a series of tests and configuration changes helped us to come up with the idea of placing blue LED strip lights on top of the Rover. This configuration offered a simple way to analyse the position and orientation of the robot, which the students could then use to determine which course to take in the simulation.

Figuring out the solution to the orientation challenge enabled us to deliver a driving solution that tracked, monitored, and directed the Rover from one point to another.

The project was completed on time and within budget thanks to the DevReady process.

Since completion, VSSEC has used its "Mission to Mars" simulation to educate tens of thousands of students. Collectively, those students have engaged in almost 1,000 missions to the Red Planet's surface. Better yet, the programs we helped to develop have empowered VSSEC to commercialise its solution. In addition to drawing more students to VSSEC, it can also use these programs to appeal to remote schools and students, providing them with a service they wouldn't otherwise be able to access.

The Mission to Mars was a success! And it all started with one Big Idea.

Coming Up With the Big Idea

What is a Big Idea?

We look at a Big Idea as a concept that you hope will shake up your niche and offer tremendous value to your consumers. With the "Mission to Mars" project, the Big Idea was to come up with a simulation that engaged students in a complex aspect of science, thus guiding them towards a career in one of the STEM fields.

Your Big Idea doesn't have to be as lofty or complex. In fact, many Big Ideas are simple ways to solve problems that amarket has. It's not the size of the potential solution that makes a Big Idea. It's the potential that idea has to have a marked impact on the market that makes it so big. A Big Idea is also something that you can take beyond your own business, as it becomes something that allows you to open yourself up to a broader market and potentially create new revenue streams. Again, we see that with the "Mission to Mars" and how it's able to operate as a remote project that is accessible to students outside Victoria.

Simply put, a Big Idea is a concept that will make a difference.

But to understand what that difference could be, you need to figure out where the massive value in the concept comes from. So, you have to understand what a great product means to you, as a business owner or project manager. You also have to understand the implications behind building this great product, specifically in terms of what you need to consider to ensure the Big Idea delivers something that your market wants and can engage with.

You may have many ideas floating around inside your head.

The Big Ideas are the ones that the market will accept to the point where the dea becomes something that has a tangible benefit on your business and the people you serve.

Evaluate and Analyse the Idea

What do start-ups and small business owners typically do?

We find that the entrepreneurs behind these types of businesses have a ton of ideas. With the best of intentions, they tend to grab one of these ideas and drive out into the fog with it. They have no clarity and can't see where they're going with the idea. They just keep driving, with their vision of the path ahead completely obscured by the fog.

After a long time spent driving, they eventually see a couple of headlights approaching out of the fog.

SMASH!

They crash the car and the idea goes up in smoke. And from there, the only course of action is for the entrepreneur to dust themselves off, grab onto another idea, and drive right back into the fog with it.

That's the scenario that you want to avoid when coming up with your Big Idea.

Instead of driving off into the fog, we want to create clarity that clears the fog. Doing this allows us to see what the journey ahead looks like and, crucially, means that we can also see any potential hazards that we might crash into.

To clear the fog, we've got to evaluate and analyse the idea. And to do that, there are six elements we need to consider.

Element #1 – Is the Market Demand Big Enough?

Is it possible to get all of the toothpaste out of a tube?

If we asked 100 people, 98% of them would say that it isn't. The other 2% might have solved the toothpaste tube problem, or at least have an idea of how they could solve it.

That's great for the 2%.

However, the question we'd have to ask here is whether the inability to get every last drop of toothpaste out of a tube is such a significant problem that people are clamouring for a solution.

The reality is that it isn't.

Most people don't really care about getting all of the toothpaste out. As a result, any idea focused on how to get all of the toothpaste out of the tube is doomed to fail because the market just doesn't care enough about the problem.

For an idea to become a Big Idea, the market demand has to be significant enough that the concept will make a sizable impact on the market.

Element #2 – Can Your Proposed Solution Solve the Problem?

An idea isn't of much use if it results in a solution that doesn't solve the problem.

Critical here is that we see the concept of solving through the eyes of our potential consumers. You may believe your idea provides a solution. However, the idea won't go anywhere if your market doesn't feel like your eventual product provides the solution that *they're looking for.*

Element #3 – Are There Any Competitors Working on a Similar Idea (Who Will Release Before You)?

Encountering competition isn't areas onto automatically shelve a Big Idea Competition is just a fact of business as there will always be companies out there looking to solve the same problems that you're focused on.

The challenge comes when there is a competitor out there who has a similar Big Idea. When that happens, you may end up in a race to get your product out before they do. This race creates pressure, which can lead to mistakes in the design and development process. It also opens up the possibility of you investing a lot of time and money into a project that still gets shelved if the competitor releases ahead of you.

Ideally, we want our Big Ideas to be disruptors that change the market, rather than concepts that end up looking like imitators because a competitor got there first.

Element #4 – Learning From the Experiences of Your Potential Customers

There's a classic quote that we often refer to when speaking to clients:

"There are too many products, too many companies, and too much marketing."

This quote comes from a research project undertaken in 1971, which was the early period of big marketing spending. And the quote reveals that a lot of customers felt like there was too much noise surrounding the products that companies were trying to push on them.

As we write this book, it's 50 years after that quote was made.

Today, there's even more noise in the market. There area lot more products, which means there are a ton of businesses out there trying to market their solutions to audiences. If you're not careful, your Big Idea could be yet another piece of noise that gets lost in the cacophony that your consumers get subjected to every single day.

That's a situation we want to avoid.

To do that, you've got to learn from the experiences of your potential customers. Listen to what they're saying about any solutions that are currently on the market. Focus on framing your Big Idea with a sense of purpose that cuts through the noise and speaks directly to your consumers.

Element #5 – What Does Your Idea Need For Youto Be Able to Sell it to Yourself?

Once you've come up with a Big Idea, you'll probably find that you can think of a million and one things that you could throw into it. However, you don't want to put everything you think of into the idea. That's a sure-fire way to go over budget while over-engineering your solution.

Instead, ask yourself what the idea needs for you to be able to sell it to yourself.

This helps you to boil your Big Idea down to its key features and benefits. You essentially create a starting point that you can iterate from over time.

Element #6 – What Does Success Look Like to your Customers?

In addition to asking yourself what it would take to sell your Big Idea to you, also ask what your customers really want from the product you're thinking about creating. What does success look like to them?

This is a crucial question because your idea of what success means to a customer may be completely different from what your customers actually want from a solution. Consider talking about your Big Idea with the people who might potentially buy it. Ask them what they think about the idea and, crucially, what they would want the eventual product to offer in terms of benefits.

What are the customer's success criteria?

Get clear on that before you start designing the product that forms from your Big Idea.

Develop Your Ideation Process

One of the biggest challenges that many organisations face when creating a digital solution is that they don't have a consistent way of approaching the flow of their project. In other words, they don't have aprocess in place for taking something from an idea, through to creating a business justification for the idea, and then turning that idea into an internal project.

Instead, organisations often start a project with the desire to make something. They build the product first before trying to figure out how it will fit into the market. The Big Idea comes and they jump straight in to the design and development, rather than putting that idea through a process that will help them to see if it has any legs.

This lack of process is one of the things that DevReady helps you to avoid. Our frame work ensures that you fully examine your idea before moving into the costly actions of designing and developing it. The goal here is to take the Big

Idea in your head and either figure out that it won't work for your market or turn it into something that you've defined and can start working on.

That's where the ideation process helps.

The elements we've just spoken about essentially act as the critical questions you need to ask about your Big Idea before you start putting some real time and effort into it. However, you also need to turn those questions into a defined and consistent process that you can apply to every single idea that you come up with.

What does that process look like?

These are some of the steps that will help you to create a consistent ideation process.

Step # 1 – The Spark of an Idea

This is the really simple step. Someone in your business, be it you or a member of your team, comes up with the spark of an idea. There's a concept floating around that may be capable of having an impact on your market. Take that spark of an idea and start building on it. Pull your team together and hold a super lightweight conversation about the idea and the potential you see in it.

It's here where you want to establish if the idea could have legs for your business. You're not thinking about how you'll build the solution yet. This is just about examining the idea and confirming that it's something that's worth working on.

Step #2 – Find the Problem

So, the idea looks like it has legs.

Now, you need to ask yourself what problem it solves for the market. Think back to our toothpaste analogy from earlier in the chapter. An idea that solves a problem that nobody cares about isn't going to evolve into a viable business proposition. The solution you offer has to be for an underlying problem that your market *needs* to solve.

Interestingly, this process may lead to you throwing away your original concept. Or, you may build on it to the point where it's almost unrecognisable. That's because the original idea is really just a spark that starts a conversation that could take you in all sorts of directions. As you delve deeper into the problem you're trying to solve, you may find that you come up with more ideas for developing a product that's cheaper, faster, or offers functionality that you hadn't thought of originally.

Step # 3 - Avoid Feature Creep

The above point about how you may build on your idea does come with its challenges. Over engineering is always a potential problem when you're developing an idea into a business case. For example, you may discover that your potential market faces several similar problems that it wants to deal with. The temptation after making this discovery is to try and design a solution that solves *every* problem that you find. In doing that, the Big Idea that sparked the process morphs into some behemoth of a concept that tries to do everything.

And that's where a lot of projects find the root of their eventual failure.

Feature creep occurs, which means you try to stuff too many features into a single project. As the features add up, so too does the time required to implement everything, in addition to the cost of taking those features from ideation through to development.

So, you need to avoid feature creep.

Doing this has to be a part of your ideation process. After you've figured out the problems and started building on your idea, sit down with your team and figure out where you need to draw the line in terms of what the product should do. What features does it need to have right now and what can't wait until the product is on the market?

By asking these questions at the ideation phase, you ensure you don't end up stumbling into a situation where you're trying to do a million and one things that delay the project. Think back to Element #5 of the evaluation and analysis period and you'll see how that element fits into your ideation process.

You Have the Idea... What Comes Next?

Coming up with a Big Idea is always exciting for a business. You have this concept that you believe could have a massive impact on your market. Of course, this also means that the idea could reap huge benefits for your business.

However, not all Big Ideas have stable business cases behind them.

We haven't started building the business case for your idea just yet. Instead, this chapter has focused on how you can conduct some of the early evaluation and analysis of the idea to see if it's worth pursuing a little further. We've revealed some of the key questions you need to ask during the early ideation process, with these questions helping you to figure out if the idea has any legs.

If it does, you're ready to move to the next step in the DevReady process.

The idea has legs and you've built on it with the help of your team. Next, you're going to do a deep dive into the potential market. You're going to investigate your Big Idea and find out if the market's telling you that it's an idea that they're ready to embrace.

CHAPTER

5

Stage Two – Investigate

During one of the early episodes of the DevReady podcast, we had the opportunity to speak to the brilliant Tim Walmsley. A retired Army Major, Tim spent 14 years in the military. He fought in both Iraq and Afghanistan before injuring his back and being forced to take a desk job within the army. Undeterred, he used the opportunity to earn a couple of master's degrees before moving into defence project management. This position allowed him to engage with government contractors and work on multi-billion-dollar projects, essentially giving him a crash course in how to conduct large-scale business operations.

It was while he was in this role that he developed the idea for Bench On, which is a sharing platform that enables businesses to generate extra revenue by loaning their staff to other businesses in need.

That business is an impressive success born from a great idea. However, Tim has also had his share of failed Big Ideas.

At one point during his military career, Tim came up with the idea for a screen door attachment. As somebody who'd spent most of his life renting, likely as a consequence of moving from place to place as part of his military duties, Tim had identified a potential solution to a problem he had. His family kept cats but they could never inst all a cat door because they lived in rented properties.

The screen door attachment was designed to enable a cat to get out of the house without putting the house itself at risk.

It sounds like a great idea, right?

Tim thought so too, which is why he poured a lot of resources into designing the prototype. However, just as he was about to move into developing the prototype, he decided that it was probably a good idea to find out whether anything similar existed on the market.

So, he jumped on Google and started investigating.

That investigation revealed that there was an Australian company, based in the same city that he lived in, that had already been making a very similar product to his for over five years.

Tim's Big Idea was dead in the water. He'd ended up wasting a month of his business life on designing a product that already existed and was on the market. However, that experience also taught him the valuable lesson that you need to conduct market research before you move into designing your product.

And that's where the second stage of the DevReady process comes in.

In this chapter, you're going to learn how to investigate the market and ensure that your Big Idea is both some thing that the market wants and also something that doesn't already exist.

Poor investigation can lead to waste, as Tim discovered with his screen door attachment. Let's find out how to ensure that your business doesn't end up wasting both time and money on a Big Idea that will never make it to market.

How to Do Market Research, the DevReady Way

What's the best way to work out what your potential customers want from a product?

Ask them.

In the first stage of the DevReady process, you've come up with a Big Idea that you believe will solve a problem that your market has. However, your beliefs about that idea are not absolute. It's possible that your market will not see the benefit that the idea has or that it wants a different solution from the one you're proposing. In some cases, as we saw with the toothbrush example in the previous chapter, the Big Idea may solve a problem that nobody cares about.

So, we need to figure out what the market cares about and what it wants from any potential solution that you create. The best way to do that is to create consumer surveys that allow you to elicit feedback directly from the people that you intend to market your product towards.

Now, building a high-impact survey is a crucial step in the DevReady process, as it allows us to gather meaningful data to inform our client's development path. As such, we conduct highly focused surveys designed to drill down into the specifics of the problems and needs within our client's niche. These surveys are tailored to reveal actionable insights that allowus to refine a product to its optimal market fit before development begins.

Here are the essential steps in designing a DevReady survey that meets these objectives:

Tip #1 – Define Clear Goals for Your Survey

A DevReady survey goes beyond collecting general feedback—it's specifically crafted to address key questions that can make or break your Big Idea. This means each question needs to align with the precise outcomes you're seeking, ensuring that you're gathering targeted insights rather than generalised opinions. Clear goals keep your survey concise and relevant, increasing the likelihood of obtaining insightful responses.

Key questions that DevReady surveys aim to answer include:

- **Does the target audience currently face the problem that your Big Idea aims to solve?**

  This question is crucial because it verifies that the issue your idea addresses is not only real but also recognised by your intended market. By verifying that the problem exists, you prevent wasted resources on a solution that solves a non-issue.

- **Is the proposed solution both relevant and appealing to the target demographic?**

  This question evaluates whether your solution is something the audience would genuinely consider. It's one thing to identify a problem, but if the market isn't interested in your proposed fix, the idea lacks viability. This question also helps pinpoint aspects of your solution that may need refining to increase appeal.

By setting these goals, the DevReady survey approach ensures that every response has a purpose, feeding into a feedback loop that continually refines your product concept. Ultimately, this step reduces the risk of pursuing ideas that are disconnected from market demand, helping you focus on solutions with clear commercial potential.

Tip # 2 – Survey the Appropriate Audience

Accurate survey data hinges on reaching the right audience. Unlike generic surveys, DevReady surveys are distributed specifically to individuals who fit your *ideal client avatar*—that is, the target profile of someone who would truly benefit from your product. This targeted approach means you're collecting responses from individuals who not only understand the problem but are also actively seeking a solution.

To survey effectively, you need to define characteristics such as:

- **Demographic attributes** (age, gender, location) that reflect your target user base.
- **Behavioral insights**, like purchasing habits or usage patterns, which indicate a genuine need for your solution.
- **Psychographic factors** that reveal attitudes, values, or lifestyles aligning with your product's purpose.

With a clear understanding of who these users are, DevReady surveys deliver more precise feedback that genuinely reflects your product's potential impact. This focus prevents skewed results that might arise from feedback collected from those who have little interest in or understanding of your solution.

Tip #3 – Tailor the Language and Format of Your Survey

How you phrase your questions and present your survey can make or break the quality of your responses. DevReady surveys are structured to communicate with clarity, using accessible, non-technical language that resonates with your target audience. Overly complex or jargon-laden questions can alienate respondents, leading to incomplete responses or even survey abandonment.

Here's how DevReady ensures clear communication:

- **Use of Simple, Direct Language:** Every question is crafted to be easily understood, regardless of the user's technical background. For example, instead of asking, "What API integrations would enhance your experience?" a DevReady survey might ask, "What other apps or tools do you wish you could connect to this product?"

- **Optimal Question Structure:** The survey avoids open-ended questions that may lead to vague responses unless deeper qualitative insights are necessary. For quantitative feedback, structured choices like multiple-choice or rating scales are used, which make it easier for users to respond quickly and accurately.

- **Adapting to User Preferences:** DevReady surveys are formatted to suit the audience's habits and preferences, whether through digital surveys, social media polls, or email formats, ensuring maximum engagement and meaningful responses.

In essence, a DevReady survey uses a conversational tone that reflects the way users naturally communicate, creating an approachable experience that encourages thoughtful and honest feedback.

Tip #4 – Develop a Structured Survey Analysis Process

Once responses are collected, the next step is to analyse the data in a way that uncovers actionable insights. The DevReady approach uses an organised framework to go beyond surface-level observations, identifying patterns and detecting specific needs or challenges that the product must address.

This structured analysis involves several stages:

- **Identifying Key Trends:** Responses are examined for common themes, such as repeated mentions of specific pain points or features. These trends help highlight areas where your product can differentiate itself.

- **Evaluating Market Readiness:** The survey analysis includes assessing the market's familiarity and comfort with potential solutions. For example, if the responses suggest that users are unfamiliar with certain technologies or functionalities, the product might need additional educational resources or a simplified user experience.

- **Prioritising Product Features:** By evaluating the aspects of the solution that matter most to respondents, DevReady can prioritise development efforts, focusing on the features with the highest potential impact. This ensures that the product design aligns directly with user expectations and needs.

- **Applying Competitive Benchmarking:** Analysis also involves comparing feedback with competitor data. If users highlight pain points that current market offerings overlook, these insights provide a unique angle for positioning the product.

Through this analysis, DevReady not only gathers data but also interprets it with a keen eye on creating actionable next steps. The process of analysing surveys within the DevReady framework is pivotal in refining the product concept, aligning it with real-world demand, and ensuring that your investment in development will result in a product that resonates with its target market.

By following these expanded, carefully tailored steps, DevReady surveys deliver actionable insights that allow your team to make well-informed development decisions. The end result is a product concept that has been validated and refined to meet both market needs and user expectations, minimising risk and maximising the likelihood of commercial success.

Understanding Your Competition

With a good market survey, you develop a deeper understanding of your audience and its attitude towards the problem you're trying to solve. You may also gain insight into your Big Idea and its suitability for solving the problem. In some scenarios, you may find that the Big Idea isn't viable, which means you have to go back to the drawing board. If this happens, you can at least take solace in the fact that you made this discovery before investing in the design and development of a product.

Of course, understanding your audience is only the first part of your investigation. As we learned from Tim Walmsley, assessing the competitive landscape is also key to understanding whether your Big Idea has legs.

You have to know what competition exists and what your competitors are doing.

As we mentioned earlier in the book, competition is a fact of business. It's extremely rare to enter into a market that nobody is attempting to serve. The odds are that there will be businesses out there that have identified the same problem you've found and are trying to solve it with their own product. Your investigation needs to bring those competitors to light, helping you to understand what they offer, how they offer it, and whether the market is currently responding to their product.

Once you have that information, you need to know what to do with it.

First, focus on the specific niche that your Big Idea serves. Do your competitors serve that same niche or are they taking a broader approach? If they're taking the broad approach, you have an advantage you can leverage in your marketing. By focusing on a specific niche, you're able to appeal more directly to an audience. Of course, you also have to ensure that the niche you're serving is large enough to make your product viable as a business opportunity.

What if a competitor is serving the same niche?

If this is the case, your next task is to figure out what makes your Big Idea different. What will your product do that the solutions that are already on the market can't do? Your product may have several similarities to what's already out there. But if you have a couple of key points of difference, and those differences allow you to serve your audience in a better way than what already exists, you have a viable product regardless of the competition.

The billionaire investor and venture capitalist Peter Thiel emphasises this point when he says a truly innovative product has to be ten times better than what's already on the market for it to succeed.

We don't necessarily agree with that.

A great product that's similar to what's already out there can still succeed if it has one or two key points of difference and is backed by a strong marketing campaign. At this stage, your task is to investigate what's already on the market and figure out if your product can offer anything different from the existing solutions. The answers to those questions then lead to an investigation of your Big Idea. Ask yourself what resources it will take to turn that idea into a product, and whether that product will be able to compete at a high enough level that you will generate a return on your investment.

What If There's a Similar Product on the Market?

Let's assume your competitor investigation reveals that another company already has a similar product to the one you're thinking about creating. Does that mean you automatically need to scrap the idea and move onto something else?

Not necessarily.

In some cases, as we saw with Tim Walmsley, the product you discover may be so similar to your idea that it has already cornered the market. In that case, it's likely best to go back to the drawing board and come up with something else. There may not be any space for another product that does the same thing, which means you could end up investing in something that gets shelved.

In other cases, you may find that a similar product doesn't actually serve your audience as well as it could. Or, you may find that the other product doesn't yet have any market traction, which means there's still room for your solution even though somebody else got there first.

So, the question now is how do you know whether your Big Idea is worth pursuing when there are similar products on the market?
There area couple of factors to consider when answering this question.

First, examine your market to see what they're getting from the existing solution. Were visit the ideas about finding a point of difference again here. By focusing on your potential client's problems, which you should understand due to your survey, you may be able to spot areas where the competitor product isn't providing the service that the client needs.

Hone in on that.

Focus on doing what you do so well that the quality of your product cannot be denied, even when compared to a similar product. Peter Thiel's advice about doing what you do ten times better than your competition is relevant here. You don't necessarily need to be ten times better. But if your similar product is better, you still stand a chance of achieving market dominance.

Second, there is the issue of price to consider.

One of the easiest ways to compete with a similar product is to offer something comparable at a lower price point. This will attract customers who prioritise price ahead of the product or service itself. However, this is sometimes a dangerous game to play. By making price your key point of difference, you commoditise your service. If another competitor enters the market at an eve lower price point, your customers will have no reason to stay with you. They'll migrate to the new competitor, just as they migrated away from your previous competitor.

By charging too little, you undervalue your service. Of course, you also have a direct and negative impact on your bottom line.

Investigation Complete

What do we learn during the Investigate stage of the DevReady process?

Through our customer surveys, we learn more about the market and its problems. This helps us to determine if we can make a business case for the Big Idea. You may discover that your market doesn't want a solution to the problem you've identified. Or, you may find that the market has other issues that you hadn't considered, which could lead to you tweaking your Big Idea.
Ultimately, the point of your market surveys is to determine whether demand exists for the product you're working on.

From there, we move onto competitor investigation. Who is already in the market? What do their products look like? How is your product different? Can you solve problems that your competitors aren't focusing on?

These are all questions you'll ask when investigating your competition. The presence of competition is not a reason to abandon a Big Idea. However, competitors who are already doing what you're trying to do, and are doing it well, may dissuade you from investing in an idea.

Ideally, the Investigation stage will reveal that your Big Idea has legs in the sense that the market needs your product. With this established, you can finally move onto designing your proposed solution.

CHAPTER

6

Stage Three – Design

We initially met Kate via the DevReady Podcast. She's the founder of A Perfect Space, which helps people to find locations for events, photoshoots, films, and tv commercials. The platform connects photographers, film producers, and event managers to stunning sites that look like they've come right out of a dream. It's as easy as choosing among the different location categories on their website and specifying all their other requirements. Add the various online tools at their disposal and everyone involved in the production can have a stress-free experience.

We interviewed Kate to learn what she'd done to build her business. And among the many valuable nuggets of information she gave us, we discovered that she was on the third iteration of her platform and that she'd spent $1.4 million so far to build what she had.

That seemed a little strangetous.

After the podcast, Kate reached out to our team.

She started asking questions about how she could achieve certain things with her developers. Again, we were curious. So, we asked Kate to sit down with us to discuss what was going on with her platform.
It was during that meeting that we discovered that her current team had not designed her platform completely. They'd built out a business requirements document.

They had some wire frames made for the platform.

But there was nothing in between! No flowcharts. No documentation. Nothing about how the platform would operate or how users would flow through it.

That's not all.

Kate knew there was something off about this whole process. Her developers had turned those wireframes around very quickly and were pushing her for approval on them. Kate was taking her time, fixing the errors she saw, and feeding back to the team. But they'd already started without her approval. And whenever feedback came in, they just adjusted what they'd built so far instead of workingwith Kate to make sure they'd gotten the design stage of her project right.

Development ended up being a bit of a mess.

Why?

The core components of design, documentation and visualisation, weren't executed properly.

Kate's developers had paid lip service to both components. They had a business requirements document and they'd built some wireframes to help Kate visualise the product. But there was so much missing in between that nobody really had a good idea of how Kate's platform would operate in practice.

The good news is that Kate caught the issue in time.

By working with us, we were able to fill in the gaps in the design, resulting in a platform that actually did what she (and her users) needed it to do.

But what if you don't get this right? What if you don't realise there's a problem with the design element of your project?

If that happens, you risk project failure as you keep throwing money at the project to fix it, rather than getting it right the first time. You'll keep spending, convinced you'll get it right next time, only to find that a poorly-handled design stage can sink your project.

And that brings us nicely to the subject of this chapter. Stage Three of the DevReady Process focuses on designing your product. Being agile in the design phase is much more cost-effective than having to reengineer your solution. To help you create an effective design, we're going to examine the two core components of design that we spoke about earlier:

1. Planning
2. Visualisation

Part One - Planning

Why is planning so crucial in any project?

The answer is that planning helps you to bring an idea to early life. Through planning, you start to flesh your idea out to see if it's viable for your intended market. And it supports its development by highlighting potential issues and providing a road map that takes the development from Point A to Point B.

And yet, so many developers speed through this crucial aspect of the design process.

They jump straight into the visualisation of the project. Or, worse yet, they start developing a project with very little planning at all.

That's how you end up with a project that requires you to keep throwing money at it. Without a clear plan , you have no direction. And with no direction, you'll never know when you've developed something that you can release to the market.

Planning your project is a critical part of the design process. There are four key things you need to do to ensure your plan is up to the task.

Step #1 - Come Up With a Minimum Viable Product

You have your Big Idea and you've investigated the market to the point where you're pretty confident that you know what it wants and that your idea for a product is probably viable.

Now, you must figure out what that product needs to have, at the bare minimum, to meet the need you've identified in the market. This is what we call the "Minimum Viable Product" or MVP.

Your MVP is all about picking out the initial problem that you've discovered and coming up with a solution for it that you feel will work for your customer base. That solution needs to deliver value. It needs to be validated. And it needs to offer enough to the user to help them see that your platform is worth sticking around for.

Your MVP *is not* a comprehensive solution to every single problem that you've identified. It's also not a version of your product that tries to do everything that you believe your product has the potential to do.

It's a version of your product that solves one fundamental problem.

With that solution, you can start building a customer base. You can validate your Big Idea and build up from there.

Step #2 – Figure Out How the Product Flows

The design stage is all about exploring.

You're fleshing out a concept, a Big Idea, here, which means you need to give yourself some room to experiment and figure things out. You've got to explore the product space, immersing yourself in the problem so you can come up with solutions.

Eventually, ideas will come to you. Designs for prototypes will start flowing. And when that happens, there is just one key question you have to be able to answer:

Does the flow make sense?

When acustomer starts using your product or platform, what do they do to get from having a problem to finding your solution? Does every step in the process you've come up with make sense? Or, are there blips along the way that could trip people up and prevent them from achieving the solution they came to your platform to get?

It's at this stage of planning that you start building out the language of your platform. You begin looking at the order of the steps you want users to take in addition to how you present information to them.

Here, you're looking at your prototype product and asking yourself if it all makes sense.

Are the buttons in the right place?

Can a user easily see how to get from one point to the next?

Can they navigate through the platform easily?

Friction is what you're really looking for here. You want to seek out the little points of friction that drag you back from achieving what you want with the product. Seek out the friction, document it, and then figure out what you can do to overcome it so that the product flows.

Step #3 - Determine the Required Processes for the Product

Let's return to the concept of an MVP.

In developing your MVP, you've got to ask yourself about the vital processesfor your product. What does it absolutely need to beable to do to provide what the customer would expect from it?

Here, you start referring back to the customer surveys you did in StageTwo of the DevReady process. You begin looking at the feedback you've already gathered so you can see what is really important to your potential customers The things those customers highlight become the fundamental processes and functions of your product.

Then, you star tasking questions about those processes.

You begin digging into how much it's going to cost to put these processes together into a solution.

Is the MVP viable if you develop all the required processes?

That last question is critical. Many business owners get so locked into designing and developing their products that they don't stop to ask themselves if what they're building is actually affordable. While they can prove that the processes and product work, they may spend so much time and money getting to a working product that they over look the fact that this product needs to go to market.

Your customers have a limit on how much they're willing to spend on your solution.

You've got to keep that limit in mind when you're designing the early version of your product. By constantly adding processes, you push yourself closer to, and perhaps even past, that limit. Strip away any processes that aren't absolutely essential to the operation of the product.

Do you still have a viable product when you only develop all the required processes?

If you do, that means you can carry on building out your design. If you can't make it work when you only have the vital processes, your design process may be telling you that the product isn't feasible right now. If this occurs, what you should do next is iterating on the processes. Consider other options to deliver your design.

Step #4 – Determine the Vital Features

Building on the topic of viability, you also need to figure out the product's vital features. Again, this comes back to our concept of MVP in the sense that you must offer at least the bare minimum features to make your product usable. However, you're also considering how vital your features are in terms of ensuring you don't overcomplicate your product.

This is something that the brilliant Miriana Taflaga, who is a human-centred design specialist, spoke to us about during one of the DevReady Podcasts. Miriana points out that you should always look at anything you design through the lens of your consumer. And when it comes to features, there is one thing that your consumer wants:

Simplicity.

As Miriana put sit:

"Simplicity is really important because people do get confused by many, many different features. And the features might be great, but combined, they are an overwhelming experience."

That feeling of overwhelm is precisely what you're trying to avoid, especially with the first iteration of your platform. If a customer lands on your product and can't figure out what they're supposed to do with it or how they can make it work, they're going to leave and likely never come back.

So, when we talk about vital features, we mean only those that solve the critical problem you identified during the first two stages of the DevReady process. If a feature doesn't move the user closer to solving that problem, it doesn't need to be in your product.

At least, not yet!

Other features can come later, once your platform is more established and you're focused on how to scale it. But during these early stages, when you're still designing the product, it's best to focus on one or two outstanding features that you can use to wow your audience as you solve their problem.

The alternative is to come up with dozens of suboptimal features that you try to fit together into some clunky product that lacks focus.

There's one final point to make here.

As you're designing your vital feature set, you may start asking yourself where the point of difference is. It's possible that your vital features aren't all that different from what's already out on the market.

This issue of differentiation is a valid concern.

But it's not a product-stopping one.

If you can't differentiate based on your features, you may be able to differentiate based on the flow you create and how those features fit together. You can improve the user experience, or by targeting a niche - you can stand out based on how you market and service the product. The point is that there are opportunities to differentiate beyond the features, so don't get too disheartened if your vital feature set doesn't seem to be setting your product apart at this early stage.

Part Two –Visualisation

In documenting your Big Idea, you start to see how all of the nuts and bolts might fit together. You focus on product flow and come up with the key processes and features that have to be in place for you to create an MVP that serves your market's needs while offering you the potential to generate revenue.

The challenge is that having the nuts and bolts isn't enough for you to figure out if the product will work.

Visualising what you intend to build allows you to see how it all fits together. That brings us to the second part of the design stage, which is all about figuring out what your product might look like in practice.

Here are the steps to follow:

Step #1 – Create a User Flow

You have an idea of the general product flow.

Now, it's time to solidify that flow into a visual format that shows exactly how a user moves through your product. Creating a flow outline allows you to do this:

Step #1 – Pick an Entry Point

Define the user's entry point, which is what leads to them using your product. For a mobile app, the entry point maybe the App Store or Google Play. For a SaaS product, it may be via a consultation. Physical products usually have their entry points in stores.

Step #2 – Create Multiple Flowcharts

Build flowcharts that take the user from the entry point to completion, with completion being the solving of the problem that you identified earlier in the

DevReady Process. It's likely that you will build multiple flowcharts here, with each defining a separate system flow. It's also possible that you may build several flow charts for the same system flow, especially if you have a few different ideas about how a user may move to completion.

The key here is that you're not defining every tiny little process that takes place at this stage. You're building flowcharts for all of the required processes that take your user from start to end. There may be a single process for this. Or, there may be several. Trying to build in every little detail at this stage leads to you getting bogged down in minutiae, rather than focusing on the main steps and processes the user must go through to reach completion.

Instead, you're defining the key steps the user will follow, as well as any choices they make along the way. Chart the course from the entry point to the final interaction.

Step #3 - Refine the Flowcharts

Once you have some draft flowcharts drawn up, you'll refine them in several ways. The first is also the most obvious–go through each of your flowcharts to ensure that all of the steps make sense. If there are any points of friction, any steps that shouldn't be there, or even any little human errors you've made along the way, this is where you fix them.

When you feel comfortable that your flowcharts collectively represent all of the interactions and journeys your customers go on, refine them to the point where they're easy for others to read. Use colours to highlight important actions, almost like a coding system that helps the reader see the information they need quickly. Maintain consistency in the use of shapes, lines, and other elements. Ensure that every label in the flowchart makes sense and carries meaning.

With that, you under stand how a user moves through your platform, at least at the MVP stage. At this point, you'll likely have multiple flowcharts, each of which represents a different process, interaction, or stage of the user's journey. You're going to use these flowcharts to develop a visual representation of the route a user takes through your platform.

Step #2 – Create a User Journey Map

What's the difference between a user flow chart and a user journey map?

The flowchart demonstrates the physical actions a user takes that move them through the platform. A user journey map helps you visualise the user's relationship with the platform, and how it evolves, during their time using it. In short, user journey maps support you in seeing your product from the user's point of view.

You can create user journey maps for every interaction within your platform. But it's best to start with a high-level map that covers the experience from end to end. Developing that map involves doing the following:

1. Figure out who your user is and what problems they have. Of course, you've already done this in the first two stages of the DevReady Process.

2. Define the scenario and the user's expectations for the journey. The scenario is what the platform is supposed to do. The expectation is what the user expects from the platform. For example, Uber's scenario is that it allows people to order rides via its mobile app. The user expects that this ride turns up on time, in good condition, and at the price agreed upon via the app.

3. List the touchpoints the user will have with your platform. A touchpoint is any interaction somebody has with what you've created. Obviously, this covers everything the user will do when on the platform. But it also covers marketing touch points, such as visits to blogs or interactions with ads.

4. Figure out what your user's intention is when they're on your platform. Ask yourself what motivates them, what their pain points are, and what actions they're willing to take to achieve their intention.

5. Create a storyboard that details each step of a user's interaction with your product. This story board may start with the user experiencing the problem your platform solves. It will then cover how they hear about your product, any touchpoints they're likely to hit before using it, and then how your product creates a change for them. Don't just focus on the steps here. Detail the emotional aspects so you understand

how the user interacts and, as importantly, why they'll continue on the journey.

With your storyboard created, you have a user journey map.

Now, let's look at an example. The following is Spotify's user journey map:

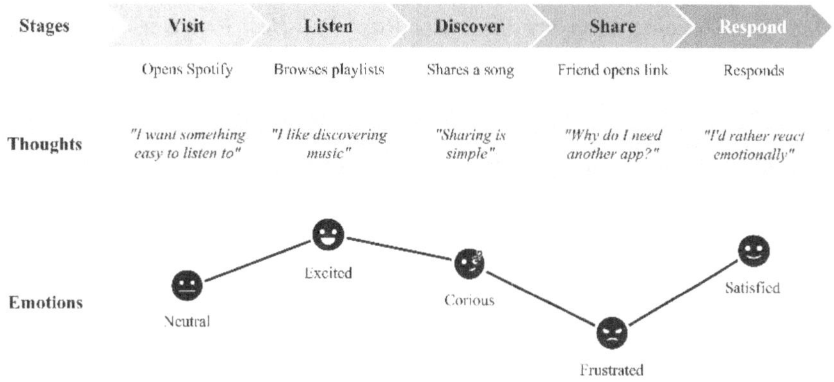

Figure 1: A user journey map illustrates how a user's actions, thoughts, and emotions evolve across each stage of interacting with a platform.

Note that this journey starts with the stages that the user goes through. It's likely that Spotify has flowcharts drawn up for every single one of these individual stages. They likely also have the steps outlined for each.

From there, the map moves into what the user is thinking and feeling as they go through each stage. It's here where the company gets to grips with its users' intentions and what they hope to get out of the platform.

Finally, we have touchpoints. In Spotify's case, they're not just accounting for the touchpoints that apply to the original user. They're also thinking about how that user can create further touch points for other users. They map both the interactions with the Spotify platform and interactions users have *about* the platform outside of Spotify.

Pull everything together and you have a cohesive map that shows how a user experiences a platform and, as importantly, how they feel, think, and talk about the platform during the journey.

Our final note is that this map is not static.

As your product evolves, so does the user journey. In fact, you may find that the actual user journey ends up differing from the one you originally designed. If this happens, revisit your user journey map, alter it, and then implement it into the design and development of your product.

Step #3 – Develop the User Interface Based on Best Practices

You've probably gathered that the visualisation phase is really all about the user and their experiences with your platform. Ultimately, your goal is to createa product that solves the user's problem while being easy for them to use.

The product's interface is critical.

If a user can't figure out what to do and how to navigate through your product, they'll just move onto something that is easier for them to interact with. Keeping this in mind, start drawing out some diagrams of your interface while following these best practices for interface design:

- Keep the interface as simple as possible all of the time. Avoid any unnecessary elements that may clutter the design while ensuring every element the user might interact with is clearly labelled.

- Consistency creates clarity in a user interface. Don't be afraid of using common elements in your design. For example, anybody who's used a mobile website has likely seen the icon with three horizontal lines that they can tap to get to the navigation bar:

There's no need to change something like that in your product. The user is already familiar with the icon so stay consistent with it. There are many similar icons defined in both Apple's and Google's design guidelines.

To find out more, see Apple's design guidelines and Google's guidance on mobile-friendly websites and platforms.

For Google's guidance on what constitutes a mobile-friendly website or platform.

- Place elements where they both draw attention and are at their most useful.

- Colour, texture, and typography can all be used to show a user how they should move through your platform. For example, different font sizes can indicate when a user has reached a new section. Colours are useful for drawing the eye towards important elements of the interface.

- Ensure you're always communicating with the user. When somebody executes anaction on your platform, there needs to be a follow-up that tells them the action was successful (or unsuccessful) and what it has achieved. Don't leave the user fumbling around in the dark trying to figure out if they're interacting with your product in the right way.

Focus on removing the friction wherever you find it. A well-defined user interface makes it easy for people to solve their problems. And if solving the problem is easy, you'll end up with a lot of happy users on your hands.

Step #4 - Define the User Experience Based on Best Practices

If your user interface designs define how the user interacts with your product, your user experience diagrams focus on actions and emotions. There's a similar relationship going on between user interfaces and user experiences as we see between user flow and user journey maps.

One focuses on the practicalities.

The other focuses on the user and how they feel.

When defining your user experience, there are two key practices to stick to:

Practice #1 - Consider the Actions

What actions would a user *want* to take on your app?

This is a crucial question to ask because we often get lost in trying to dictate our users' actions when designing our products. Instead of designing based on what we think the user should do, we design based on what the user wants our platforms to achieve.

There's a definite distinction between these two things.

For example, you may have a product that helps users to deal with a highly personal or embarrassing problem. If that's the case, is the user really going to want social media sharing buttons plastered all over that product? Or, are they going to expect a discrete experience that helps them to solve an issue that they're not comfortable sharing with other people?

Actions aren't about what you think the user should do. They're about what the user *wants* to do.

Practice #2 – Focus on Emotions

How do you feel as you move through your designs? Do the colours convey the right tone? Does the language uplift you or help you to connect with your pain and your desire to find a solution?

These are all questions you must ask when designing a user experience.

For example, let's say you're developing a mobile app for a health spa.

Heavy music, dark colours, and harsh language would all create an emotional experience that doesn't align with what the spa wants to achieve. To create that emotional alignment, you'll use soft and muted colours along with tranquil music.

How your users feel is just as important as what they do.

If the atmosphere you create within your product doesn't align with what the user wants to achieve, that user will leave!

Step #5 – Build Wireframes

Do you remember MirianaTaflaga from earlier in the chapter?

She has a simple piece of advice when it comes to visualising your product:

"Draw wire frames and put words together."

Wireframes are low-fidelity drawings you use to represent your product, its interface, and the journey that your users take with the product. As you've followed these five steps, you've likely already started creating some basic diagrams, particularly when figuring out your user interface.

With wireframes, you bring all of your design ideas together. Think of these diagrams as being the equivalent of the blueprints an architect creates for a building. The better the blueprint, the easier it is for the builders to create a great product. The same goes for the solution you want to develop. With strong wireframes, you define what your product is supposed to achieve while offering guidance to help your developers put it all together.

It seems like drawing wireframes should be the easiest thing in the world. But there are a few practices we recommend as part of the DevReady Process to ensure your wireframes do what they're supposed to:

- Be clear about what the wireframe is supposed to achieve. A lack of a clear goal leads to a directionless sketching process.

- The first sketch is not the final sketch. Though wireframes are basic representations of your design, they still need refining into something useable.

- Flow is important so use arrows and lines to show how one wireframe drawing links to the next. Remember that you're using wireframes to depict an entire experience rather than just a basic design.

- Functionality is always more important than aesthetics when it comes to wireframes. You can always spruce a basic design up. But if the underlying processes don't work, it won't matter how pretty the design is.

- Annotate and edit liberally because wireframes are not things you need to get attached to. Accept when changes must take place and recognise that wireframes are just tools that help you to make your final product better.

Wire frames are more than basic pen and pencil sketches on scraps of paper. They're the blueprints of what you want your product to be. And though these blueprints may change over time, you still have to put effort into them so they can aid you in clarifying your thinking and designing your product.

Does the Design Fit the Bill?

In conclusion, the design stage is a critical part of the DevReady process.

Yet so many people skip over it or only give it half-hearted focus. Even those who recognise its importance, like Kate of A Perfect Space, may still get caught out by working with developers who don't understand or respect the two core components of quality design.

It's not just about creating wireframes or crafting the perfect user interface—it's about truly understanding your users, their needs, and the emotions they experience as they navigate your platform. By following the steps outlined in this chapter, from mapping the user journey to refining your user interface and experience, you're setting the stage for a product that is not only functional but also intuitive and aligned with your users' expectations.

However, there's another crucial step here: validating your wireframes and design with stakeholders before moving into development. This process goes beyond securing buy-in; it's about ensuring that the solution you're building truly aligns with the needs and expectations of every one involved. Stakeholders, whether internal team members or external users, bring valuable perspectives to the table. They can point out gaps, raise concerns, and identify potential roadblocks early on, which can save you time, money, and frustration later in the development phase.

When validation is done right, it refines your design, clarifies the development process, and helps to avoid costly mistakes further down the line. This iterative, collaborative approach fosters transparency, strengthens alignment, and ultimately leads to a product that has been vetted by the key players who will ensure its success.

By treating the design stage as a collaborative, ongoing process—where feedback is not just welcomed but actively sought after—you're building a much stronger foundation for your product. You're not just checking boxes; you're shaping the future of your product in a way that's thoughtful, user-centered, and aligned with both user needs and business goals. This approach significantly increases the chances of your product's success, and by continuously refining your design before jumping into development, you're setting yourself up for smoother, more efficient progress towards a final, polished product.

CHAPTER

7

Stage Four – Estimate

In 1995, University of Kentucky researcher Albert L. Lederer teamed up with University of Dayton researcher Jayesh Prasad to examine what was then an emerging problem in the business world:

The cost of digitising services.

Together, they created a paper called *Causes of Inaccurate Software Development Cost Estimates*. As part of their research, they spoke to leaders at 112 organisations, discovering that cost estimation for information systems was a key concern for the vast majority. The duo highlighted that underestimating the costs of a software project leads to drastically inflated budgets, sometimes even resulting in the project's cancellation. Companies end up pouring good money after bad, desperately trying to salvage a project until they reach a point where they can do no more. Then, they have to accept defeat, along with the knowledge that they've spent millions of dollars trying to develop something that will never see the light of day.

It's a surprisingly common issue.

In fact, Lederer and Prasad's research led them to statistics from a Dallas consulting firm that estimated that 70% of its clients ended up spending more than they'd estimated on their most recent software projects. The paper also highlighted a Business week article that examined an insurance system that ballooned from an $8 million cost estimate to an actual development cost of $100 million.

Again, remember that this was in1995.

Just imagine how much more that actual development cost could have been if that system was developed today.

We're sharing this story with you for two reasons.

The first is to highlight that issues with cost estimation are nothing new. For as long as we've had technology, we've had to overcome issues relating to our estimates for how we'll be able to leverage that technology. As part of designing and developing your product, you're going to have to estimate costs for everything from development through to the physical components you'll need.

Sadly, the estimate will likely be inaccurate if you don't follow a good process for creating it.

That brings us to the second reason why we're sharing this story.

In their study, Lederer and Prasad point out that a lack of understanding of, or changes to, project requirements are not the chief cause of budget overruns. If anything, they're the symptoms of problems that can betraced back to the management level:

"In reply to the long-standing question as to whether overruns are more attributable to inaccurate estimating or ineffectual project management, the data in this study raise concerns about the latter. The results suggest that information systems managers and professionals might want to reassess their attitudes toward their users, and researchers might want to focus on particular causes of inaccurate estimating."

The point?

Project management has a huge impact on your ability to estimate a budget. If you don't have a process in place to create cost estimates, you're likely to overrun. If you don't understand the various factors that go into creating an estimate, you're likely to miss something that causes you to go over your budget.

Spending more than you budgeted can have a disastrous effect on your project.

For a large business, the extra money spent is money that can't go towards other projects or to shareholders and owners. For start-ups, the effects are more dramatic. Poor cost estimation canlead to the failure of a project and the closure of the business that's tied to it.

This chapter is all about ensuring that poor cost estimation doesn't lead to the downfall of your product. In Stage Four of the DevReady Process, we estimate costs effectively by understanding the factors that impact costand what you may be able to do to bring your project costs down.

The Digital Transformation Challenge

In a 2019 article for Forbes, AI influencer and podcast host Michael Gale shared some interesting insights into the concept of digital change. He points out that the vast majority of business leaders know that digital change is a thing they need to do. Most leaders recognise that evolving with the times, both in terms of how their businesses operate and the products they offer to people, is essential to survival and the only way that a business can thrive in the 21stcentury.

And yet, 50% of business owners are still dragging their feet when it comes to digital transformation. They're not taking the required steps to bring their businesses and products into the modern world.

That's shocking enough.

But the real take away from the article is not that half of business owners aren't following through on the need for digital transformation. The takeaway is that the vast majority of those who do follow through end up *failing*!

In fact, according to Gale, only 28% of businesses that attempt digital transformation do it successfully. Meanwhile, the other 72% end up making 62 cents less for every dollar that the 28% make. We're starting to see the emergence of a business world that contains digital thrivers and digital failures.

Most fall into the digital failures category.

Right now, you might be asking yourself why we're sharing this information with you. After all, you may be the head of a tech company. Even if you're not, the fact that you're reading this book means that you understand the need for digital transformation and are currently trying to figure out what it takes to achieve that transformation or, at the very least, to create a product that satiates the modern market.

The reason we share this is simple.

Of the 72% of businesses that fail to achieve their digital transformations, many are large companies with millions, if not billions, in their coffers. They're able to throw a lot of money at their projects. And yet, they still end up with results that generate meagre, if any, returns.

What does this tell us?

How much money you invest is not the critical component in bringing a piece of technology to life. What's far more important is how you think when designing and developing that product. The DevReady Process focuses on changing how you think so that you're able to design a product that is both ready for the current market and will fit within your budget.

Money is obviously important. This entire chapter is dedicated to ensuring that you don't spend too much by helping you to create more accurate estimates. The crucial point we want to make here is that throwing money at the platform or project you're trying to develop is not a solution. You can't move into development under the assumption that you can solve any problems that occur by spending more money on them.

By accurately estimating your costs, you also gain clarity on the true scope of your project and what it will take to bring the project to life. Understanding that scope means there's less chance of overspending. It also means there's a higher probability of you creating a product that reaches the market.

Why Cost Estimates Can Vary Wildly

Way back during the second episode of the DevReady Podcast, we had the opportunity to interview Mellonie Francis. She is the founder of Find Book Beauty, which she describes as being the Airbnb for beauty services. Mellonie's platform connects beauty professionals with their potential customers.
It's a great idea.

Mellonie used Squarespace, along with a few plugins, to develop her basic MVP. And in less than 12 months, she had what was essentially a proof of concept that her idea was going to work. With the MVP proven, her next step was to create a custom platform that moved her away from Squarespace. She wanted complete control over Find Book Beauty. Plus, she didn't want to have to keep paying a third-party provider money to keep her platform running.

Eventually, Mellonie accepted that she'd need to have somebody develop the solution for her. And that's when she began running into hurdles. She got intouch with several software engineers and web designers to get their estimates on how much they thought her project would cost to develop.

The numbers ranged from $60,000 to $500,000!

There was so much variation in the estimates that Mellonie didn't know which numbers she could trust. Were the estimates at the lower end of the scale missing important aspects of the project that would create additional cost later on? Were the estimates at the higher end of the scale far higher than they needed to be based on the project's scope?

Mellonie didn't know.

And this is a situation that we see so many non-tech people get into when they're trying to figure out how much it will cost to develop their products. The estimates can vary so wildly that they may feel like they've been plucked out of thin air.

How do you figure out an accurate costestimate for your project?

The first step is to understand the factors that can affect the estimate that you or a developer may come up with. There are eight, each of which has an impact on the eventual cost.

Factor #1 – The Way the Platform is Built

How does the developer build their platforms?

Some developers create customised platforms from scratch. They pour hours into building solutions that offer everything you want to incorporate. However, all of this extra work comes at a cost. Building everything from the ground up takes a combination of time and talented individuals, which leads to costs climbing very quickly.

Other developers find ways to keep costs down while still offering a quality deliverable. For example, the developer may build a cheap system using WordPress rather than custom-building a web platform. They, or you, may use low-code or no-code tools, such as Bubble. io, to get the barebones of a platform in place that you can scale up later. Going this route allows you to speed up development at the cost of customisation.

Ultimately, your choice here comes down to what you want to provide as your MVP. If you want to build a quick MVP, such as a WordPress installation with plugins, that you'll throw out later, the open source route may be the best choice.

But if you want to build a stronger MVP for the first stage of a launch or because you want to attract funding, it may be worth investing more in a customised platform.

Factor #2 - Customor Templated Design?

When Mellonie decided to move away from Squarespace, she did so because she didn't want Find Book Beauty to have a templated look. She wanted a unique and bespoke platform that did exactly what her users needed it to do.

You may want the same for your platform.

But if that's the case, you have to accept that a custom design will cause the cost of your project to shoot up. It takes time and effort to build something that's brand-new. Developers and designers need to collaborate closely to ensure every feature, user flow, and branding element is perfectly aligned with your vision. This level of precision and flexibility is invaluable if you're looking to create something that stands out or meets specific functionality requirements, but it comes at a premium.

By contrast, using a template is often as simple as plugging in content and tweaking colour schemes and branding. This approach can be much more cost-effective and quicker to implement, making it a great choice for projects with limited budgets or simpler requirements. However, the trade-off is that templates may lack the depth of customisation needed for highly unique user experiences or complex features.

The decision between custom and templated design ultimately depends on your project's goals and priorities. Are you looking for a one-of-a-kind solution that fits your needs like a glove? Or can you achieve your objectives with a solid, cost-effective starting point? The DevReady process helps you navigate this choice by aligning your budget, timeline, and vision to the most appropriate design.

Factor #3 – Where Is Your Team?

The location of your development team has a huge impact on your development costs. For example, the average annual salary of a software engineer in Australia is $116,880. That amounts to $59.94 per hour of work. An offshore software engineer in India may cost between $19 and $40 per hour. If you're working on a multi-year project, that saving of $20 or more per hour may be what you need to continue investing into the platform you're creating. However, you're saving that money by working with an engineer who is in a different time zone and who is not directly attached to your company.

Ultimately, you have some pros and cons to weigh up here.

Competency is a key concern. With an onshore developer, you're able to meet directly and assess the developer's skills before you hire them. You likely won't have that opportunity with offshore developers, meaning you have to rely on a portfolio of work to make your decisions.

Differences in practices is also a key concern. As we can see from the above stats, hiring an offshore developer will likely work out to be cheaper per hour than hiring on shore. But what if the overseas team calculates 50 hours for a task that a local team would take 10 hours to complete with your direct oversight. Now, you're not just losing the savings you thought you were going to make. Your offshore team is actively costing more money because it's taking longer to complete the tasks assigned.

There are other relative factors to consider.

Would you rather spend less on an inexperienced offshore team or more on an experienced local developer?

Is an offshore team capable of developing a product that is of the quality needed?

Will language barriers lead to a lack of clarity that could result in costly miscommunications and delays?

On a general level, having an onshore team costs more, but it allows you to directly oversee development while working closely with the team. Offshore teams cost much less per hour and often provide access to some amazing talent. Unfortunately, you may struggle to manage your project when the team isn't available directly or when you're trying to juggle work across different time zones. Once you consider all of the relative factors that play into your decision, you may find that the offshore option is costlier than you anticipated.

You may be able to find a happy medium by hiring an onshore project manager whose sole duty is to oversee an offshore team. Regardless, your choices here have a huge impact on your cost estimates.

Factor #4 - Freelancer or Agency?

Let's assume you don't have an in-house development team. That means you're going to have to work with an external company.

Do you go with a freelancer or an agency?

The agency will cost a lot more. However, you'll also pay for the experience of an entire team and everything that brings with it. A freelancer will usually cost much less on an hourly basis. However, you may find that they can't cover the full scope of the project. Still, a freelancer may be able to do the job if the product only requires minimal coding work.

Again, you must assess your project needs to figure out which is the right option for you.

Factor #5 - The External Company's Model

It's not just your own business model you have to consider when cost estimating.

You also have to account for the models of any companies you work with. This need to account for other companies' business models and working methods is what led to Mellonie receiving such varied estimates for the development works he needed. The general rule is that the larger the agency, the larger the quote you can expect. However, this isn't a hard and fast rule.

Smaller teams tend to offer greater agility and flexibility in their work. However, it's possible that a large project will spread a small team too thin, resulting in the team struggling to provide the expected level of quality.

Furthermore, leaner agencies will likely focus on a single project at a time. While they may not spend the money on meeting you in person as often as you may like, they will usually be focused on your project alone, rather than having several projects running at once.

Large agencies tend to have multiple management layers. They may invest time and money into flying representatives around the country to meet with clients. They'll also maintain several teams so they can take on more projects. All of these points add to the cost of working with them.

However, large companies also have more human capital, which means they have larger pools of talented and experienced team members who may be able to save you money in the long run. Where as mall team may take a year to complete a task, a large company with more experienced members may get the job done much faster.

There is no "correct" choice here.

The point we're making is that you have to consider the model of any external company that you work with. How will that model affect you from a cost perspective? What are the pros and cons of the model an external company uses? Would you prefer the flexibility and focus of a smaller team? Or, does the collective experience of a large agency mean you're more confident that you'll get what you need?

These are all questions you have to ask when choosing who to work with.

Factor #6 – Project Specifications

Here, we dig into the project management aspect of cost estimates that Lederer and Prasad highlighted. If the estimates you're receiving vary wildly, or the estimate you've created doesn't seem to make sense, there's a high probability that the estimate doesn't cover all of the project specifications.

These variances in estimates are more likely during the early stages of design. As you develop a more solid idea of what your product needs to do, so too should you be able to more accurately estimate the cost of development. The more clarity you establish in the design space, the higher the chances of accurate estimation.

Ultimately, you must understand your product specifications to estimate accurately. The first three stages of the DevReady Process are designed to give you that level of understanding.

Factor #7 – The Talent in a Team

Earlier, we shared some statistics about onshore and offshore development costs. We compared the average rate in Australia, which is nearly $60 per hour, to the rates in India, which can vary from $19 to $40 per hour. The cost-sensitive among you would likely gravitate to the Indian developers in this instance. However, it's crucial to note that higher rates are not necessarily a bad thing and low rates are not always a good thing. There's an element of "you get what you pay for" to consider.

Talent costs money.

If you're intent on working with the best developers, you have to expect they will charge a premium for their time. But you'll know that you're getting good work and that it will likely be completed on time and within the agreed-upon budget.

Going for the cheaper option typically means that you're working with a developer who may be less talented or is currently unproven. This may not be a bad thing, depending on the project's complexity. If you only need some fairly simple work done, working with the crème-de-la-crème just means you're paying more for something you could have gotten a lot cheaper elsewhere. However, going for the cheaper option can also lead to delays and the delivery of a product that isn't what you anticipated it to be.

Factor #8 – What You Want Out of the Platform

Adding to all of this is what you want to achieve with the platform you're developing. The decisions you make at the early stages often play a huge role in determining the cost estimate for a deliverable.

Do you want a system that you can keep building over time? Or, does your platform need to be complete from the moment you launch?

Does the system have to be able to scale with you?

Could you modify open source technology to build your product?

Are you happy with a vanilla system, with one database and one web view, in the early stages? Or, are you willing to pay extras to develop a more complex system?

Ultimately, you're asking how "complete" your deliverable has to be for it to work as an MVP. The answers to these questions influence every choice you make in terms of how your platform is built and who you work with during development.

The Two Things You Need To Understand When Estimating

We can see how the decisions we make have an impact on our cost estimates. The choice between developing in-house, onshore, or offshore impacts cost. So do choices related to agencies, freelancers, and the talent levels of those on the project. Even the models employed by the businesses that you work with have an impact on how much your project will cost to deliver.

But there are two more things you need to wrap your head around beyond the factors we've already covered:

1. Your Project Timeline

2. The Project's Resource Requirement

Let's take a look at each issue.

Issue #1 - Your Project Timeline

Every project you undertake has its own issues and complexities that you, as a project manager, must deal with. Successful project management involves balancing all of those unique requirements so you can create a timeline for delivering the project.

The larger the MVP, the longer the time frame and the more that time costs. More time means more functions can be added to your MVP, though this again leads to you spending more money on wages, contractors, and other project-related expenses.

As we said, it's a balancing act.

Ideally, you're looking to crunch your MVP down into smaller pieces that you can work on individually. This reduces the level of risk you face because you're no longer working on a large and complex project. The larger an MVP gets, the riskier it gets. If you can break it down, you reduce your timelines and, in many cases, lower development costs.

Before you start development, figure out how long you *intend* for the development process to take. This involves doing the following:

1. Determine the size and complexity of the project. Hopefully, you've done this during Stage Three of the DevReady Process, meaning you should have some wireframes and design documentation drawn up. The better your understanding of the project's scope, the more likely you are to create an accurate estimate for how long it will take to bring to fruition.

2. What approvals might you require before you can move forward? If you're working within a large company, you may need approval from stakeholders or other authoritative people to move forward. Or, you may be working in a highly regulated industry, which means you need to get approval on whether your project is actually viable from a legal standpoint. How long is it going to take to secure permission to move forward? That time gets added to the project timeline.

3. Create a contingency plan for when something goes wrong. In a perfect world, this will mean coming up with a detailed set of steps to follow if something threatens the established timeline. For example, what

will you do if your developer has an accident and can't work on the project for an extended period? Create backup plans and think about how having to implement them could impact the project timeline.

Considering and accounting for these factors early on will shorten your timeline and, ideally, lessen downtime if things go wrong. In both cases, that means money saved that you can account for in your cost estimates.

Issue #2 – The Resource Requirement

You need resources to complete a project.

That much is obvious to anybody who's building a physical product. In this case, materials and the equipment needed to build the product are some of there sources you've got to account for when estimating cost. You must also consider the cost of the labour required to bring the product to life, in addition to the cost of transporting your product to stores or customers.

Those looking to develop software platforms have resources of their own to consider. Again, there's the people side of things. You can't program a software package if you don't have programmers. Add to that the design and testing portions of software development and you have some pretty extensive resource requirements. Then, there's the need for specialist software or technology to create your software.

Again, those technologies are resources that your estimate must account for.

All of this means that you need a way to determine what resources you need so you can allocate them effectively in your project. These quick steps will help you:

1. Review available resources and compare them against your project's design documentation and visualisations.

2. Identify your currently available resources so you can figure out where the resource gaps lie.

3. Review any organisational policies around resource usage. This is more of an issue for those working within businesses than it is for

those who own their own companies or who are exploring their ideas independently.

4. Analyse alternatives wherever possible. For example, you may have a web designer in-house who can only dedicate one day per week to your platform. Perhaps hiring an offshore designer who can work five days per week on the project is a better option?

5. Make your resource purchasing and allocating decisions based on the tasks required and the identified resources. You may find that breaking some more complex tasks down further helps you to quantify them.

By confronting the issues of timelines and resource allocation, you figure out what your project needs, when it needs it, and, most importantly, how these requirements affect the cost.

How You May Be Able To Cover Some of the Cost of R&D

We've spent a lot of this chapter talking about costs and how to account for them.

Let's switch focus to something a little more positive – how you can cover some of the costs of the research and development work needed to bring your project to life.

In an early episode of our podcast, we hosted Caitlin and Alex of R&D 360. The company specialises in helping people and businesses gain access to tax incentives that can help them push their projects forward.

They informed us about the many opportunities that exist to claim tax incentives for research and development (R&D) work, particularly in Australia. However, they were also quick to point out that accessing these opportunities requires you to do more than present a project to the appropriate authorities. In most cases, you need to have a "core activity" that you can present as the basis of the project that you're making the claim for.

In your case, the Big Idea you hope to bring to life will likely be your "core activity".

However, not every Big Idea will qualify as an appropriate core activity. You need to be able to prove that there is risk in the project. You must establish that you're doing something novel, be that in terms of your approach to the project or the technological components required. A "core activity" has to be technically challenging. If your Big Idea involves using a lot of pre-existing technology and ideas, it's unlikely that you will be able to claim the tax incentives Caitlin and Alex highlight.

With the core activity established, you may be able to make claims against the cost of that activity, thus lowering your tax bill and covering some of the costs of the research and development work that is so essential to your project.

They also pointed out that you may be able to make claims for supporting research and development activities. Examples might include testing, documentation, background research, team meetings, project management activities, and anything else that you do to support the core activity.

There are many potential examples.

But the general rule of thumb is that if an activity is necessary to undertake to support your core activity, you can likely claim against it as a support activity. This all applies to Australia. And sticking with Australia, there are several programs in place to help innovative companies attract funding to cover the costs of R&D. The most prominent of these is the "Research and Development Tax Incentive" (R&DTI).

As of 1 July 2021, the R&D tax offset is structured as a premium applied on top of an entity's corporate tax rate, with variations depending on turnover and R&D intensity.

For R&D entities with aggregated turnover of **less than $20 million,** the refundable R&D tax offset is calculated as the entity's corporate tax rate plus an 18.5% premium.

For R&D entities with aggregated turnover of **$20 million or more,** the non-refundable R&D tax offset is based on an incremental premium linked to R&D Intensity.

- **For eligible R&D expenditure up to 2% R&D Intensity,** the non-refundable tax offset equals the corporate tax rate plus an 8.5% premium.

- **For additional eligible R&D expenditure above 2% R&D Intensity**, the non-refundable tax offset is the corporate tax rate plus a 16.5% premium.

These changes replace the previous tax offset rates that applied up until 30 June 2021, which included:

- A **43.5% refundable tax offset** for companies with annual turnover of less than $20 million.

- A **38.5% non-refundable tax offset** for companies with an annual turnover of $20 million or more.

Companies engaging in eligible R&D activities should review the latest guidelines provided by the Australian Taxation Office (ATO) to ensure compliance and maximise their claims under the R&DTI program.

Our point?

There is help available.

You don't have to cover the entire cost of your project alone. And if you know where to look, you can find grants, tax breaks, and other incentives that are designed to support innovators like you.

We've focused on Australia here.

However, there are similar programs available all over the world.

The United States has the Research and Development Tax Credit, which uses a set of tests while assessing how much you've spent on R&D to provide a tax credit that essentially covers some of your costs.

The United Kingdom also has similar programs, including its Research and Development Expenditure Credit and its Small and Medium-Sized Enterprises R&D Relief Scheme.

Again, the point here is that you can find ways to cut your costs so you can lower your estimates. Even if you're not in a position to attract investor funding, there are schemes, often government-backed, that provide support and tax incentives for the creators to go out and do what they do best.

Take advantage of any schemes that you can find. It may also be worth speaking to a tax professional who specialises in working with companies that have R&D requirements to ensure you're getting access to all of the tax breaks you can possibly have.

Make Sure You Get Your Estimate Right

Cost estimation is a crucial part of developing a new product.

It's also the part that so many businesses and innovators get wrong. And the simple fact is that people get it wrong because it's difficult to estimate the costs involved in a project. It's this difficulty that leads to so many varying quotes from suppliers. Unfortunately, it's also this challenge that leads to such a huge number of projects going over budget.

So, what can you do to ensure you're creating more accurate cost estimates?

It all comes down to clarity.

You need to be clear of the project requirements and exactly what they entail. If you've worked through the first three stages of the DevReady Process, you already know what you're trying to achieve. More importantly, you have documentation and wireframes that make it easier for any potential partners to see what you're trying to build.

You must also be clear on the many factors that can affect a cost estimate. Everything from how you allocate resources to the specific type of agency you work with can impact the cost of your project. Some of the most critical factors to consider at this stage include:

- The size and scope of your project.
- The team you'll work with.
- Your approach to the technology you'll use and how you're going to build the project.

Many developers will find they can bring their costs down by using off-the-shelf products when establishing their MVP. Often, those that try to do everything from scratch are also the ones who find themselves mired in expensive and long-winded development processes.

Accuracy is essential at this stage of the DevReady Process.

The more accurate your cost estimates are, the lower the likelihood of you getting stung by unexpected costs later on.

With your estimates established, you're a step closer to beginning development. However, there are still some more steps to take, starting with the validation of your project.

CHAPTER

8

Stage Five – Validate

Your first idea isn't always your best idea.

This seems like another obvious concept. However, we see time and again that innovators get so locked into their initial ideas that they don't see the issues inherent in those ideas. At times, these issues can be such that the entire project gets placed at risk.

We have the perfect example for you.

We work with an engineering group based in Melbourne. We've also spent a lot of time implementing the DevReady Process with them. One day, the company came to us with an idea.

They had identified that the contracts they give to employees were often complex to the point where the employee struggled to get to the meat of the contract. Their idea was to "cartoonise" the contracts. They wanted to create fully automated cartoons using a fully customised platform.

It was an interesting idea.

And seeing as they came to us to develop it, we could have told them we'd go right ahead and start trying to implement their idea.

But that's not what we did.

We took it back to our offices to validate the idea. We ran through it with a fine-toothed comb and found that taking the approach of having fully dynamic cartoon processing would lead to the creation of hundreds, if not thousands, of iterations of cartoons. All of which would need to be automated. The client would also likely have to go down the 3D route in creating the cartoons.

The costs started adding up.

What started as an idea to create some little caricatures ended up being a platform with a cost estimate of between $100,000 and $200,000!

That wasn't viable for the client.

So, we advised them to take a look at some off-the-shelf products rather than developing a product from the ground up. Doing so would lower costs considerably. We also asked them to consider using preset cartoon characters, rather than the dynamic and customisable option. Again, this would reduce complexity, leading to lower costs. In stead of trying to go the fully-customised route, check what's out there already and see if you can integrate it into your existing systems.

The client did just that.

And they ended up using an off-the-shelf product that allowed them to develop two characters rather than dynamic cartoon characters for far less than the six-figure sums they would have spent on a customised solution. The HR team even ended up winning an award for innovation in the engineering space.

As we said, the first idea isn't always the best.

In this case, the first idea was a fully customised cartoon creation platform that would have cost hundreds of thousands of dollars to set up and maintain. The concept of using cartoons to explain contracts wasn't a bad one. It's just that the initial idea for a solution proved unviable.

How did we know the idea wasn't viable?

We validated it.

We took the idea, explored it from every single angle, and helped our clients gain complete clarity on what the concept really entailed.

And that is what Stage Five of the DevReady Process is all about. Now that you've documented your design and have a cost estimate ready, it's time to validate your Big Idea to ensure nothing is standing in your way of making it a reality.

The Stages of Validation

Validation is a four-stage process that includes the following phases:

- User Testing
- StakeholderValidation
- Timing Validation
- Business Case Validation

Each stage helps you to create clarity in your project, ensuring you know what you're getting yourself in for in the development stages while ensuring you feel confident that your Big Idea is something that the market and stakeholders will support.

Stage #1 – User Testing

One of the best ways to validate a project is to get the people who will use what you're developing to, well...use it!

User testing lets you see how people use your product. What do they do when they have it in their hands? Do they follow instructions? Do they go off and do their own thing? Is there anything the user does that might break the product or disrupt the flow of how they move through it?

These are just some of the many questions you'll be able to answer with user testing. The point of this phase is to observe what happens when you set users loose on the product. Of course, you haven't designed a complete product at this stage. Instead, you likely have a prototype or wireframes created based on your design documentation. In preparing that prototype, you made a lot of assumptions about how people would use it.

User testing either validates or shatters those assumptions.

It may show you that the assumptions you've made about your target market's problems are off base. You may discover that the real problem is not the one you're focusing on. And that discovery could change the direction of your project or, in some cases, cause you to scrap the whole thing and start again.

So, how do you ensure your user testing gives you accurate outcomes?

Being careful of the questions you ask is important. When we're tied into an idea, it's easy to allow our biases to influence the questions we ask of our users. If you're not careful, you may end up leading users towards the answers you *want* them to give, rather than allowing them to guide you towards what they need from your product.

Instead, keep your questions more open-ended.

Ask them what they like and dislike about your solution. Why do they like or dislike these things? What was their experience like? Was it useful or not? Did it solve their problems? What would they change?

These are all questions that cause the user to truly think about the solution you put in front of them. There are also questions that encourage answers that allow you to see where your product is working and where it may need changes to ensure it's offering a solution to the main problem you identified earlier in the DevReady Process.

Stage #2 – Stakeholder Validation

If you're not working alone, there are likely several stakeholders attached to any project you are working on. It is important to bring your stakeholders on the design journey to increase stakeholder buy in and also give you a greater chance of a valid and viable product.

Take the engineering company that we spoke about at the opening of this chapter. We worked primarily with the company's HR department. For them, the stake holders would be the people within that department who are responsible for implementing solutions. There are likely also stakeholders in senior management positions that the HR department reports to. And the employees signing the contract are also stakeholders.

All of these stakeholders have opinions on what is and isn't feasible for a project. What's more, they often have control over access to funds and other things that influence the success of the project. If stakeholders don't validate your idea, you're likely going back to the drawing board.

This is a concept that we touched on during our podcast with Miriana Taflaga. You may remember her as the human-centred design expert that we introduced in Chapter Six when talking about determining the vital features of your product.

Miriana recommends speaking to stakeholders regularly to ensure they buy into the idea that you're working on:

"So, you could bring in all your stakeholders and say, 'Okay, we've got this idea concept. We know the customers have this problem element.' So, you've covered that off. And then you actually have really honest discussions about what's viable and what's not."

Those conversations ultimately help you to see what your stakeholders think about the project. Those stakeholders can then tell you what's viable based on what's happening in their areas. For example, somebody who is responsible for budgeting may tell you that your cost estimate is too high, resulting in you going back to see what you can do to reduce the estimate. Or, somebody in R&D may point out that the tech your project needs is either inaccessible or not currently available.

Again, the idea goes back to the drawing board.

The main goal of stake holder validation is to provide the project's key stakeholders with the information they need so they can make a decision.

That's it.

As somebody who's seeking validation for their project, you may find yourself feeling tempted to skew this process. Again, that's a sign that you're so connected to your idea that you may be unwilling to hear from people who raise legitimate issues. Stakeholder validation is a time for honesty. It's a time for you to put all of your cards on the table so you show your stakeholders what you have.

From there, it's up to the stakeholders to weigh your idea, and the information you've presented, against the economic, environmental, safety, and capacity issues related to the business.

When you have project stakeholders, they're responsible for the final decision about whether a project moves forward or not.

It's your job to provide them with the information they require to determine whether your project is useful, usable, capable of solving the problem it claims to solve, and, perhaps most importantly, commercially viable.

Stage #3 – Timing Validation

We all know that Apple's iPad is a huge success story. The product has been through several iterations, sold millions, and inspired a slew of imitators.

It's also nowhere near close to being the first tablet computer introduced to the market.

Over a decade before the introduction of the iPad, Microsoft created a tablet computer that Bill Gates believed would become the most popular form of personal computer in the United States within five years of its release.

That didn't happen.

Instead, Microsoft's tablet computer disappeared soon after entering the market.

On the surface, Microsoft's tablet seemed to have a lot going for it. The device used the same Windows XP operating system that was popular in PCs of the time. It also came with a handy little pen, meaning it has some touchscreen functionality. The Microsoft tablet was like having a PC in the palms of your hands.

Yet it still failed. Why?

Timing.

Apple has proven that the tablet idea was a good one. But perhaps the biggest difference between Apple's iPad and Microsoft's tablet is that Apple's product arrived at the right time. The iPad arrived in an era where we'd gotten used to having little computers in our hands thanks to smartphones. It also came at a time when Apple could sell the product for between $499 and $699.

Compare that to Microsoft's tablet, which retailed at $2,000 ten years before, and you can see how the cost was a huge restriction for the company. It was also an unavoidable restriction given that Microsoft could only work with the technology of the day.

The simple point we're making with this example is that the world wasn't ready for tablet computers when Microsoft released theirs. It would take almost a decade of technological advancement for that idea to become viable both in execution and as a commercial opportunity.

Of course, this is far from the only example of a product being ahead of its time.

Take video calling technology as another example.

Today, making video calls is normal. We have Facebook and WhatsApp allowing us to make video calls for free over the web. Applications like Zoom even allow for huge video conferences, all using the combined power of clever software and the internet.

However, the concept for video calls existed long before these companies made them a mundane reality of life. In fact, AT&T even went so far as to develop multiple video calling products throughout the 20thcentury.

Their first was 1964's Picture phone.

Unveiled at Disneyland California, the product did exactly what it claimed to do–allow people to make video calls. But it never caught on. And neither did the many other attempts AT&T made to introduce similar products in the years that followed.

Why?

The technology hadn't caught up to the idea yet. AT&T even said themselves that their video calling products were too expensive, too intrusive, and, frankly, unable to provide the level of quality that users hoped for.

The introduction of webcams, followed by cameras in phones and laptops, eventually made the idea of video calling viable.

AT&T was too ahead of the times with their idea.

We're sharing these examples with you because timing needs to be a very conscious part of your decision-making process. Is the world ready for your idea?

Is the technology you have access to capable of supporting the idea while allowing it to be commercially viable?

Does consumer behaviour show you that the market is ready for the idea?

Are culture and society ready for it?

These are all questions you must ask before moving forward with your project. Failure to get the timing right can result in thousands, if not millions, of dollars being spent on a product that the market just isn't ready for.

Stage #4 – Business Case Validation

A business case is a set of documents you create to demonstrate why a product is needed and what benefits it will bring to both the company and the market when it's introduced. Your business case is what you'll present to the major decision-makers and the financiers of your project.

If you're looking for investment, it is a good idea to build a business case.

If you're working within a large company, the key decision-makers in that business will examine your business case before giving you the go-ahead.

These people either validate or reject your business case based on what they see in your documentation. Fortunately, much of the work you've done in the DevReady process up to this point serves as the foundation for your business case.
You have your Big Idea and you've investigated to confirm that the idea is something that the market has an interest in. You've also worked through the design stage, meaning you have written and visual representations of what

you believe your product should be. You even have a cost estimate that takes a range of factors into account.

All of this helps you to demonstrate the inherent value in your idea while also showing that it's something that can be achieved within a certain cost parameter.

Now, you just need to arrange that information in such away that it clearly communicates the value to anybody who's validating your project. There are five steps to follow if you wish to create an effective business case.

Step #1 – Outline the Opportunity

Place the opportunity your project represents front and centre. Talk about how the product helps solve a problem worth solving for your target customer. But more importantly, highlight how the idea creates profit potential.

Step #2 - Discuss the Options

Consider the various approaches you could take during development. For example, you could stack the outsourced approach up against working with an in-house development team. The key here is to show that you've taken the time to gather information about every potential approach, which increases confidence in your eventual selection.

Step #3 – Evaluate YourOptions

You've outlined the potential approaches to the project. Here, you evaluate how (orif) each of those approaches can deliver on the project's objectives.

Consider both the risks involved and the financial value each approach creates.

It's here where you start digging into return on investment. Everything about an approach may seem to make sense on paper. The cost for development may seem right. The idea may look like it will work. However, if you can't demonstrate a clear path from idea to return on the investment into that idea, you may not have a valid business case for the approach you're considering.

Step #4 – The Implementation Strategy

Your evaluation leads to the selection of the approach you want to take with your project. Here, you talk about how you're going to implement that approach. What resources will you need? What milestones will you set? What does the product look like and what are you doing to mitigate risks? You'll answer all of those questions here.

Step #5 – Your Recommendation

Your business case concludes with your recommendation of the approach you believe should be taken. Understand that those you present this business case to may not agree with your recommendations.

Obviously, you'll look to external parties to provide validation of your business case. If an investor or lender says "no" after seeing your business case, that suggests there is a problem that may benefit from a second look. However, it also helps to see the creation of your business case as a means of validating your preferred approach. You're forced to consider other options when creating a good business case, which means you might find a different approach works better than the one you thought would workbest.

It's not just about getting validation from others.

It's about validating your approach in your own mind.

Is Your Concept Valid?

The point of the Validate stage of the DevReady Process is to confirm that your Big Idea is something that the market wants and that you're capableof delivering.

Your user testing helps you to see what the market is getting out of your product. It also aids you in refining your designs and prototypes to ensure the product delivers as intended.

Conversations with stakeholders allow you to consider issues that may not have been apparent to you when coming up with the idea. They also represent a chance to get key stakeholders on your side, though you need to ensure you're honest through the entire process.

Timing is the least tangible form of validation in this stage of the DevReady Process. You're essentially validating that you have the tools to create and deliver your product at an affordablec ost, in anacceptable format, and to a market that's ready to accept it.

Finally, the creation of your business case allows you to validate your approach to yourself, as well as validate what you're doing to those who might finance your project.

Once you're confident that you have a valid and viable idea, you're yet another stepcloser to the development stage. But before we move to Stage Six of the DevReady Process, let's stop and talk about your tech team.

CHAPTER

9

Stage Six – The Tech Team

Have you ever watched Formula One racing?

On the surface, it seems pretty straightforward. A driver gets into a high-performance car and races it around a track for about 70 laps. Of course, the skill involved is far from simple. But what you see on your TV screen doesn't take a lot of effort to explain.

Formula One is a solo sport where the best driver wins.

Except, that's not the case.

Behind every Formula One driver – and their car – is a pit crew that must be switched on every second they're in action. Each crew member needs to be so in-tune with their specific job so they can execute it in seconds, with a level of accuracy that is simply breathtaking to watch.

Just look at the average Formula One pit stop.

It takes less than three seconds for the crew to change a vehicle's tires, make whatever adjustments they need, and refuel the car for the rest of the race. In less time than it's taken you to read this chapter up to this point, a pit crew can complete a complex job at a high speed, while under a tremendous amount of stress. To make that happen, every crew member (along with the driver) must trust that each of the other members knows precisely how to do their job. The pit crew members need to communicate effectively with each other and follow a defined process that gets the car ready to take on the rest of the race in a matter of seconds.

So…

Why are we talking about Formula One pit crews in a book about developing digital solutions?

The simple answer is that these pit crews are outstanding examples of high-performance teams that can execute at the highest level consistently. They're so talented at what they do that Formula One racers literally trust them with their lives. After all, if something goes wrong with the car, it'll careen off the track and potentially injure the driver inside.

We're bringing up Formula One pit crews because they offer the perfect example of what a great team looks like in action. What's more, we think that you can learn an awful lot from these pit crews when building your company's tech team.

It doesn't matter that pit crews and tech teams are different. What matters is that pit crews demonstrate all the qualities that make great teams work. In this chapter, we will dig into those qualities in more detail before explaining how you can build a tech team that is every bit as effective as a Formula One pit crew.

What Qualities Does Your Tech Team Need to Have?

We all have vague ideas about the qualities that a great tech team needs to have. You're probably thinking about many standards, such as the ability to communicate and a clear goal.

Formula One pit crews have those qualities too.

But they shine in specific areas that your team needs to emulate. By focusing on the following qualities, you'll develop a high-performance team that feels unstoppable.

Quality No. 1 – Every Member Knows Their Role

Take a minute to watch a pit stop in action. Slow the footage down so you can see it all. There will be about eight or 10 people surrounding the car when it pulls in. And every single person has a job.

There's the guy who raises the car so that everybody else can do their work. There's somebody positioned at each wheel to unbolt the tire. As soon as the tire's loose, somebody removes it while somebody else slots the new tire in place so it's ready to be secured. You'll also see other people around the car making rapid adjustments focused on a specific area.

But do you know what you won't see?

Crossover!

The guy who unbolts a tire doesn't follow that up by carrying it away and getting a new tire.

That's not their role.

Instead, they stay in place, ready to do their job again when the new tire is in front of them.

What about the guy who lifts the car?

That one stays precisely where they are, watching the rest of the pit crew do their work so they know exactly when to drop the car back down again.

The point is that every pit crew member understands what role they play, and they don't diverge from the role for a second.

Each member needs a defined role that enables complete focus on the task at hand. You need to encourage that same laser-like focus within your tech team. Don't overburden your people with so many different roles that they don't know what to do first.

Quality No. 2 – There Is No Star Performer

Nobody stands above anybody else in a pit crew. Everybody has an important job, with everybody else recognising how crucial that job is. Even the driver plays a role in the team, as they're responsible for pulling in accurately and pulling out again as soon as the rest of the team finishes.

There's no favouritism.

Each team member understands how critical every other member is to the team's success. That's an essential quality for a tech team because relying on a star performer only works until something happens to stop that superstar from doing their job.

Your brilliant coder isn't much use if the systems go down, right?

Quality No. 3 – Prioritisation

Pit crews don't try to do everything at once. A Formula One car may have several technical issues, with some being more important than others. A good team knows how to prioritise so that it handles significant problems first and leaves the more minor issues for another time.

In other words, your tech team needs to understand prioritisation.

By limiting what each member does so they're only working on the important stuff, you prevent them from getting side-tracked by work that doesn't matter right now.

Quality No. 4 –Preparation

Watch some more footage of a Formula One pitstop.

Notice how every pit crew member is in position long before the car arrives. That preparation means each member can jump right into their assigned task without messing around. There's nobody in a panic trying to rush into position. Everybody knows what's coming, which means they're ready to do the job. The lesson here is that preparation is essential.

A tech team that rushes into design and development work without preparing properly is almost sure to fail. And even if the team doesn't fail, the lack of preparation means resources aren't spent optimally, leading to cost inefficiencies and time delays.

As the old saying goes, you need to sharpen your axe before you can chop down a tree. In other words, your tech team needs to prepare appropriately before it can complete an important task. This need for preparation is exactly why ideate and investigate are the first twosteps of the DevReady process.

Quality No. 5 – Perfect Flow

Goals and processes.

These are two vital components of getting into a flow state. And when you watch a Formula One pit crew, you see both in abundance.

The goal is simple–get the car ready to get back on the track.

The processes are far more complex as there are several operating simultaneously to reach the goal. We've already discussed a bunch of them. What's important here is that the processes are so nailed down that there's absolutely no delay. Pit stops are examples of fluid execution built from processes that each team member has mastered.

By the way, that mastery is the other component required for flow.

By combining a shared goal with appropriate processes and the mastery each individual brings to the table, a team can operate fluidly to achieve its objective.

Quality No. 6 – Failure Hurts

What happens if a member of a Formula One pit crew fails to follow the process and complete their role.

Pain.

Potentially a lot of pain. For example, a loose tire could cause the vehicle to veer off as soon as it pulls away, placing the driver and everybody near the car at risk.

The fact that failure hurts is a constant reminder to Formula One pit crew members that they need to be process masters.

It works a little differently with tech teams.

For tech teams, failure is part of the job description. Ideas and designs go through several iterations before they become something you can use. Experimentation is key to ideation and innovation.

So, it may seem that pit crews don't have a lot to teach us here.

But that's not the case.

Failure to execute a process correctly is just as damaging to a tech team as a pit crew. By not adhering to the process in place, a tech team member could derail a project by causing delays or simply failing to implement something crucial to the project.

Failures in ideation and experimentation just lead us to new ideas. But failure in the process can lead to a good idea never seeing the light of day.

How to Pick the Right Tech Team

You know that you're looking to build an efficient pit crew when creating a tech team.

There's just one question…

How do you build that team?

Look for a High Work Ethic

We had the opportunity to speak to the wonderful Judy Celmins for the very first episode of the DevReady Podcast. In 2011, Judy started a journey. The owner of a marketing company, she had a vision for creating a piece of tech that would allow marketers to survey and question people within rooms. Think of it as a marketing-focused version of Slack. You brought people into a room, talked to them about a topic, and then used that room to conduct a survey related to the topic.

That technology went through several iterations before it was ready for market.

And perhaps the biggest challenge that Judy faced lay in finding the right team for the job.

All she was left with was a half-finished code she couldn't use. Judy called a bunch of external development companies to help, and they all said the same thing: They weren't going to touch it.

So, Judy started looking further afield. Instead of limiting her search to teams in Australia, she scoured Europe to find some software superstars to help her with her project.

And she found them in Latvia.

Some providence must have come into play here because Judy's husband is of Latvian descent.

Judy engaged the Latvian team and asked them if they could work on the project. They said yes and Judy's idea was back on!

During the podcast, we asked Judy what she saw in the Latvian team that made her pick them.

She answered that every member of the team had a stellar work ethic. And it's that work ethic that you need to look for when building your tech team. You can have the most talented coder in the world, and it isn't going to matter if they're not willing to put the effort in. Work ethic drives people to learn, grow, and become more effective as part of a team. If any tech team member isn't willing to work hard, they're probably not cut out for your team.

Confront Communication Barriers Early

Once she'd established that her Latvian team would work as hard as she needed them to, Judy had another issue to confront.

Communication.

But it wasn't the language barrier that proved to be her biggest problem. Judy freely admits to beinga bit naive when it comes to communicating in tech language. She told us that she'd often write down what she expected from the team, only to find that she hadn't written it in the style that the team needed it.

Getting past that barrier was a huge learning curve for her.

And it may be for you too, especially if you're not technically inclined. Developing a piece of software isn't as simple as you saying, "I want this," and your tech team making it happen. You need to be willing to dig into specifics. You need to spend some time working directly with your tech team so that you understand the unique language the team speaks, and they can get a feel for how you communicate.

Understanding is key.

If a tech team member isn't willing to take the time to explain what they need from you, there's a communication barrier that you need to break through.

You Have to Trust the Team You Choose

Let's switch our focus to another entrepreneur who appeared on our podcast.

Mellonie Francis is the founder of Fine Book Beauty. Just like Judy, she faced trials and tribulations when working with advisors and tech teams.

And there's one thing that Mellonie says you need to have with your tech team:

Trust.

Trust is essential, especially if you're not a technical person yourself. You must be able to trust that your team will lead you in the right direction in the project. Your team must be able (and willing) to guide you along and show you how and why certain things are done.

Without that trust, you don't have a relationship.

You'll spend too much of your time second-guessing your team because you don't believe they can do what they say. And your team won't trust you either.

The team will sense that you're always over their shoulders all the time, meaning they won't feel as comfortable communicating with you.

Collaboration is key.

Like with a pit crew, every team member needs to know their role. But just as importantly, you need to trust that the person in each role is the right person for the job. If the trust breaks down, your relationship with your team follows soon after. And more often than not, your project stalls because a lack of trust leads to bickering and a loss of focus.

Consider Bringing a Tech Advisor on Board to Help

Fear can be devastating when you're trying to build a tech team. You know you're about to pump a lot of money and resources into your project. That leads to anxiety, which can cause you to fear making decisions.

Fear is exactly what Mellonie experienced.

She felt so confused by all the frameworks and technical documents she was seeing that she was too scared to make the decisions her tech team needed her to make.

So, she brought a tech advisor on board.

That advisor helped her "decode" all the technical stuff that she had little experience with. Plus, the advisor enabled her to choose the right people for her team. Getting advice from somebody who knew what they were talking about helped Mellonie stay on the right path.

Remember what we said about pit crews earlier:

There is no superstar on the team.

At Least One of the Team's Leaders Needs to Have a Tech Background

It seems so obvious, doesn't it?

Of course, your tech team leader needs to have a tech background. Except, the leaders of many tech teams don't. They come from the business world but haven't necessarily worked with tech.

Again, this is a situation that Mellonie encountered with one of her early tech teams. The CEO of the agency she chose said all the right things. He had a track record of releasing his own products.

But he wasn't a tech guy.

And it didn't take long for Mellonie to see that he had the same communication challenges that she had when talking to the team.

The lesson here is that somebody at the top end of the team you choose needs to have some tech experience. It doesn't necessarily have to be the CEO. But somebody needs to understand enough about programming that they can understand the challenges the team faces.

A team leader with a tech background will help bridge any communication gaps between you and your team.

How to Work With Your Tech Team

You know what you're looking for in terms of qualities. Plus, what we've just shared should hopefully help you as you start building your tech team or searching for an agency that can help you turn your idea into reality.

But there's another challenge we need to confront.

Having a great tech team doesn't mean much if you don't know how to work with them! So, we will share a few tips that will help you get the most out of the time you spend with your team.

Tip No. 1 – Start by Working Together on a Project Roadmap

If your tech team doesn't understand the project's requirements, they're not going to be able to deliver on the project. It really is that simple.

Without a roadmap, your team won't know what you're trying to deliver. And if the team doesn't know what they're delivering, they won't be able to map out the steps they need to take or the milestones they need to hit. Furthermore, the team will struggle to help you get down to the root cause of the challenges affecting the project.

When a team doesn't go through the roadmap process, they deliver software based on surface-level information. The result is usually a solution that has the wrong focus.

Don't let that happen.

Before your tech team starts developing anything, sit down with them and create a project roadmap. Get involved from the beginning, so you communicate your needs effectively, allowing the tech team to figure out what it needs to do to bring your vision to life.

Tip No. 2 – Create a Communication Protocol

Judy Celmins saw how hard it is to work with a tech team when a communication barrier occurs. Only once she figured out how to break through the barrier did her team become more effective.

But let's get one thing clear here.

Communication isn't always a language issue.

Often, it's a frequency issue. If your tech team only provides updates once every couple of months, you will have a communication problem.

Communication barriers can also arise if your team doesn't tell you why an idea will not work. Teams that just plough ahead and try to implement everything you ask for without providing their own feedback can lead you in the wrong direction.

Simply put, communication is crucial.

Even if you're not technical, your tech team needs to be able to communicate properly with you. Of course, you also need to communicate effectively with them. The best solution here is to create a protocol that dictates how you communicate and when. Set up a schedule for catch-up meetings and encourage your team to share feedback and bring up challenges they're facing.

And remember this one nugget of wisdom:

If your team can't explain your project to you, how well do they actually know it?

Tip No. 3 – Don't Allow Freelancers to Run the Team

There's nothing wrong with hiring freelancers.

They can prove essential for helping you to tackle little pieces of your project. A good freelancer may bring ideas or expertise that your tech team doesn't have.

But having a freelancer lead your tech team might not be the best idea.

Why?

Freelancers split their focus. They may have full-time jobs, meaning they freelance on the side. Or they're juggling several clients at once, meaning your project never has their complete focus. To add, freelancers have limited familiarity with your team, as they may not have worked with the team before.

Any freelancers you hire are suitable for overflow work and small tasks. But unless you're able to hire them to exclusively work on your project, they can't serve as the leaders of your tech team. Ideally, you need to give those leadership roles to people working on your project and nothing else.

Tip No. 4 – Recognise and Celebrate Success

People like it when you show appreciation for the hard work they do.

It's a simple concept, and it's one that most of us try to follow when working with teams directly. But if you've hired an agency, or built an offshore team, it's often easy to forget that you need to celebrate their successes too. After all, you're not working with them directly. They're not in your office, meaning the little celebrations that represent so much can slip your mind.

You don't have to do anything too complicated here.

Just make sure to highlight successes during your meetings and conversations with your tech team. If the team cracks a milestone, talk about it. Celebrate the fact that you're a step closer to finishing your project. Showing appreciation and celebrating success creates deeper team cohesion and enhances the team's connection to you.

Tip No. 5 – Keep an Eye Out for Scope Creep

As good as your project roadmap may be, new needs will usually come to light during development. You may come up with new ideas or your tech team discovers issues that require more complex solutions than you initially thought.

It happens.

The key is that you don't allow it to happen too much. If you try to get your tech team to accommodate every new bell and whistle you think up, you'll end up with a project mired in scope creep.

If a new feature or idea gets brought to the table, work with your team to ask some questions about it.

Does the project need this idea?

How long will it take to implement, and will the idea deliver a benefit that makes it worth the additional time?

Can the idea wait until you've established your MVP in the market?

Remember that a good idea you have today can always be implemented later down the line. You don't have to release software that contains everything. By keeping an eye on scope creep, you ensure your team stays focused on the things your solution absolutely needs to have.

Get Your Team on Your Side

Spend a little more time thinking about pit crews. Watch more footage and observe teams that work in perfect harmony.

That's what you're aiming to achieve with your tech team.

You want the people in your team to know their roles so well that they can execute them at the drop of a hat. The people in your tech team need to prepare, understand what they're doing, and communicate well to ensure your project stays on track.

Start by reinforcing the qualities that your tech team needs to have. Flow – the ability to prioritise – and the ability to prepare properly are all crucial here. Once you know what you're looking for, move on to building the team. If you're struggling, consult a tech advisor who can help you make the difficult decisions. Look for people who have a high work ethic and are just as focused on breaking down communication barriers as you are.

Your work isn't finished when you have the team in place.

As the project's leader, it's your job to keep that team focused and happy. Establish communication protocols to ensure you're getting regular updates. Keep an eye on scope creep so your new ideas don't side-track your team. Celebrate success and ensure that anybody who takes a leadership role is 100% committed to your project.

And of course, create a project roadmap so that every member of the team sings from the same hymn sheet.

We've talked a lot about building a team in pretty general terms here. In the next chapter, we're going to get more specific. The success of your digital project hinges on you finding great project engineers for your team. That's what we're going to focus on next in Stage Seven of the DevReady Process.

CHAPTER

10

Stage Seven – Engineering

It was supposed to be a marvel of modern engineering.

When the Tacoma Narrows Bridge first opened to the public in 1940, it made headlines for being a cable suspension bridge and the first bridge to use plate girders to provide roadbed support. It was also the third-largest suspension bridge of all time, with its 853-metre central span.

The engineers used nearly 15,300 cubic metres of concrete and more than 270,000 kg of structural steel to create this marvel. And it cost $6 million to build, which is almost equivalent to $1 billion in today's money.

But just four months after its completion, the Tacoma Narrows Bridge collapsed.

The warning signs were there.

Observers had already noted that the bridge had a habit of swaying dangerously in windy conditions. The bridge's movement was even noticeable during normal winds, which suggested design problems that the engineers hadn't accounted for.

On the day that the bridge collapsed, the wind was reportedly moving at 110 kms per hour.

These winds created torsional vibration, or swaying, in the centre stay (or the highest vertical columns of the bridge). Over time, the swaying movement increased in amplitude until the bridge itself began twisting. The centre stay somehow remained stable, but the bridge itself kept rocking back and forth.

The failure of a cable on the north side was the culprit that led to the disaster.

The loss of that single cable set in motion a series of events that eventually pulled inwards the columns holding the cables. Huge cracks appeared on the bridge as a result. And in no time at all, the bridge was nothing more than a pile of rubble lying in the river below.

Thankfully, no lives were lost due to the Tacoma Narrows Bridge collapse.

The failure of the bridge's design provided engineers with some key learnings. That's because after creating a scale model of the bridge, researchers tested the model in high wind conditions. They found that the bridge's shape made it aerodynamically unstable, leading to the creation of that swaying movement that pushed the girders supporting the bridge out of position.

But the biggest revelation?

This was not a new challenge in bridge construction.

Engineers were aware of the existence of these issues. They'd gone unnoticed during the Tacoma Narrows Bridge construction because they were never specified as potential problems. So, what we had here was a failure in project engineering that could have led to the loss of hundreds of lives.

Why are we telling you this story?

There's a simple lesson to learn from the tale of the Tacoma Narrows Bridge:

Poor product engineering in the early stages of development can lead to major problems with the product when it goes live.

Thankfully, your project likely won't cost as much or pose as drastic a potential human danger as the Tacoma Narrows Bridge. But the unfortunate fate of this revolutionary project shows us that working out what you need from an engineering capacity is a vital step in the development process.

And to work out what you need, you need to bring great project engineers into your team.

But before you can start looking for project engineers, you need to know what engineering software is all about.

You're Building a Digital House

Engineering software is a lot like building a house.

When you start building a house, you look at the land first.

You search for a specific patch of land suitable for the building you want to create. Then, you spend some time looking at display homes to get some idea of what you want to build and learn more about the competition and what's currently going on in the market.

Then you move onto features and requirements.

How many bathrooms does the house need to have? What about bedrooms? You need to come up with some ideas for the property before speaking to an architect to create a rough layout for the property.

With the layout in place, you move onto engineering.

Your engineering team creates all the technical drawings that explain how you'll put the house together so that it'll look like what you have in the plans.

And this is the crucial stage.

Engineers handle all the technical stuff so that your builders – i. e., your developers – don't have to think about it. In the case of a house, this means creating technical drawings that specify where load-bearing posts and beams need to go. These drawings cover every aspect of the house from a technical level long before you start building it.
What do the builders need to put this house together? Where do the beams need to go? How much wood do you need for the walls?

These are all questions that get handled at the project engineering stage.

And thanks to that project engineering process, the builders just need to follow a blueprint that takes them from Point A to Point B.

It's a simple and logical process.

And yet, it's a process that many of us fail to follow when creating digital solutions.

In many cases, an entrepreneur gets an idea for something they want to create. They'll spend some time drawing up storyboards and creating a rough roadmap for the idea. But when it comes time to handle the project engineering, they skip over this vital step and jump straight into development.

They don't think about how they're going to build the project.

They just start building and hope that everything works out for the best.

For housebuilding, the equivalent would be if the builders started laying the slabs and foundations for the house without an architecture plan. They'd be building walls and door frames without a clear picture of what should go where. So, chances are, they're going to make big mistakes like setting up a door that opens right into a wall, instead of the open side.

And worst of all, they won't find out they've made a mistake until they've got the door frames up… and realise the door doesn't swing right!

This results in a lot of delays and extra costs.

And the same is true for your development process.

If you skip the engineering step, you force your development team to make things up as they go along. And without certain conditions or logic points set up before the development, they'd just be winging it, working on executing something in the middle of trying to understand what the big picture is all about.
Only when they've made costly mistakes that require them to rework the code will they realise they've been doing it wrong this whole time.

But by taking the time to get the engineering right, you ensure that your developers geta blueprint that shows them exactly what they need to do. It would be like giving them a logical architectural plan that has already accounted for all the possible paths before determining which one is best.

So, all that would be left for you and your development team to do is execute that plan as they write the code.

It means you'd no longer be thinking about *how* to solve the problem. After all, that's what the development stage is all about. At this point, you would have already solved the problem in engineering. You'd just be implementing and executing that plan.

And that's why the engineering process is so important.

It allows your team to design and plan what they'd be working on before *actually* working on it.

Sure, this 'plan' might not always be 100% bulletproof. But it will certainly leave you in a better position than being halfway through coding and only then realising that something's wrong... or that you have to do something different.

Another way to describe the engineering process is like giving your developers a recipe.

You don't want your developer just throwing ingredients into a pan and cooking it and then finding out it doesn't taste right. Because then, they'd be left redoing the dish over and over again until it tastes right. Imagine how much time, money, and effort would be wasted if you let this happen!

So, this is what you do instead:

You let the engineer think about what ingredients are needed for the dish, including how those ingredients would complement each other or how they might detract from one other. And once they're done picking out all the right ingredients, you get them to write detailed, step-by-stepi nstructions on how to cook the dish.
This way, all your developer would have to do is execute the cooking of the dish according to the recipe.

Note that in some instances, your developer and your engineer might be the same person. But this doesn't make the recipe, or the architectural plan, any less relevant. Even if it's the same person who has to plan and execute the software, it still pays to have them undergo the planning process before they jump into implementation.

The Qualities of a Great Lead Engineer

Now, you know how important the engineering process is. Your lead engineer has big shoes to fill and that is why they must be held to a high standard.

So now, let's look at the must-have qualities of a great lead engineer.

Experience is a given. We've already covered that above. It's also one of the more difficult qualities to find, which is why great lead project engineers are so rare. But if we look beyond experience, there are several other qualities that your lead engineer needs to display.

The Ability to See the Big Picture

This is where the experience factor comes into play.

Many inexperienced developers struggle to see the big picture when working on projects. Instead, they see the tasks they need to complete and little else. Developers often don't focus on the business side in terms of what they need to deliver and how the end product serves the customer's needs.

And that's okay.

Working on the coding and nothing else is the role that most of your developers play in the tech team. Your project engineer is responsible for seeing and communicating the overarching project vision to these developers.

Your lead engineer needs to be able to see the big vision. They need to look beyond the project's technical aspects and understand how the deliverables their team creates solve the problems your customers experience.

The Ability to Simplify

If your team gets lost in the complexity of the code, your project becomes overblown and more costly.

Good lead engineers do not feed into this complexity.

Instead, they look for ways to simplify wherever possible. That starts with ensuring all the code behind the solution is clean, simple, and easy to work with. Cleaner code is more understandable, making future edits and changes much smoother.

This simplification extends to the front-end delivered to customers. Project engineers should constantly ask themselves if a process that takes several steps can be simplified into something that places less of a burden on the user. If it can, the engineer steps in, formulates a solution and shows their team precisely what they need to do.

Simplicity may seem like an alien concept in the complicated world of tech development. But it's something that every project engineer should strive for.

Soft Skills

Where software developers focus on creating code, a lead engineer needs to do so much more.

In addition to having the coding expertise needed to guide their team, lead engineers must also be logical, patient, and committed. They need to understand the importance of deadlines and be able to communicate effectively with their teams. In short, great project engineers back their technical skills up with the soft skills that only come from experience working with people.

An engineer who only focuses on the project's technical aspects may get things done quickly. But they'll end up losing the support of their team if they don't have the soft skills needed to keep developers motivated and on track.

A Hands-On Approach

What happens if a project engineer does nothing more than handle the technical side and then gives orders to the tech team?

The team starts to see the engineer as this unreachable entity who sits on high and only communicates when they need something done. Even worse, the team may start to question the engineer's abilities because they see no evidence that the engineer knows what they're doing when it comes to coding.

What do project engineers do to avoid this perception?

They take a hands-on approach to their work.

Great lead engineers aren't afraid to get down into the trenches with their team. They're capable of building simple blueprints that they can explain to the rest of the team. Plus, they may occasionally get their hands dirty by showing less experienced developers how to create code more efficiently.

The Desire to Elevate Others

In the last chapter, we talked about how no single member of your tech team is a superstar. And yet, having a lead engineer seems very similar to having a superstar that the entire team relies on.

That's not necessarily the case.

Good lead engineers aren't interested in the personal glory or accolades that come with being involved in a major techproject. And they're certainly not afraid of working on less glamourous tasks for the benefit of the team.

Project engineers are teachers.

They share difficult challenges with their teams because they know the developers improve when they overcome complex problems. Project engineers adapt quickly and create learning environments that elevate the people they work with to their high standards.

Heading Over to the Most Exciting Part

With this chapter done, you now know how important the engineering process is and how it can set your development process up to succeed or fail.

That means we're going to move on to the most exciting stage of the DevReady process.

You have all the people in place. You've designed your project and are confident that you can produce a marketable solution that appeals to consumers.

Now, it's time to develop the thing!

CHAPTER

11

Stage Eight – Develop

It was the early 1600s, and the Swedish King Gustav II was at war with Poland. The conflict had raged since 1618, leading theking to conclude he needed a more substantial naval presence in the Baltic Sea to gain the advantage.

But there was a problem.

Most of the vessels in the Swedish navy were tiny patrol ships that had very few guns.

Knowing he needed to strengthen his fleet, Gustav decided to build some larger and more heavily-armed ships. Along with his advisors, the king decided that this new ship must be 108 feet long and have 32 cannons on its gundeck. The project was the most ambitious in Swedish naval history. And to make sure it was completed correctly, Gustav hired a master shipbuilder named Henrik Herbertson.

Henrik felt pretty confident in his ability to build the ships that Gustav wanted. So, he accepted the commission and told the king that he'd have his new vessels in two years.

But that's not what happened.

Even though the king needed the ships urgently, he couldn't stop himself from changing the original specifications he'd relayed to Henrik. Finally, after months of tweaking and changing, the king approved the design of a ship that he would call The Vasa.

Materials were provided and Henrik finally got down to building the ship based on his original calculations.

Tragedy struck a few months into the project. A terrible storm devastated the Swedish naval fleet, destroying 10 of its ships. Panicked by the loss of such a large number of boats, Gustav sent a simple instruction to Henrik:

Speed up!

Gustav told his shipmaker not to worry about the budget or resources. His only focus was to be the speed at which he could complete the project.

Henrik adjusted his plans so that his team could work faster.

A few more months passed before a letter arrived at Henrik's workshop. It was from Gustav and it contained instructions to increase the ship's length to 120 feet, thus allowing it to carry more military cargo.

This new instruction presented a problem.

Henrik accepted the original commission because the ship would be 108 feet long. He'd built ships of that size before and knew exactly what they needed. But a 120-foot ship was new to him. And remember that this was the 1600s. Henrik didn't have any computers or advanced mathematical tools to help him figure out how to change his original designs.

Flustered, Henrik sent a letter back to Gustav.

He told him that he couldn't possibly extend the ship beyond 111 feet in length, especially if he was to achieve the accelerated deadline for the project.

Though he wasn't pleased about this message, Gustav acquiesced and allowed Henrik to build the 111-footship. Henrik's team got back to work. A few short months later, they'd made a 111-foot keel in record time.
But the team couldn't celebrate this victory.

Another letter arrived from Gustav that talked about a larger Danish ship containing a double row of guns. Naturally, Gustav wanted his new ship to compete, so he told Henrik that the ship now needed to be 135 feet long and have 32 cannons.

That wasn't physically possible with a 111-foot ship.

Exasperated, Henrik replied that to make this new and even largership, his team would need materials dedicated to developing otherships. Gustav gave the go-ahead and Henrik's team started work on a re-design for a ship that would be larger than any they'd ever built.

Henrik's team expanded to the point where he had 400 people working day and night to extend the ship's keel.

You might guess what happens next...

Another letter from Gustav arrived a few months later.

This one told Henrik that the ship needed to hold 36 cannon rows instead of 32. And that's not all. It also needed space for 12 small cannons and the room for small munitions.

And so it went for two-and-a-half years.

As soon as King Gustav II got a new idea, he changed the project's scope. After Gustav decided the ship now needed 64 cannons, Henrik moved heaven and earth to serve his king. When the Danish implemented more attractive design elements to their ships, Gustav decided The Vasa needed to look even better.

It was that final request for decorations that proved the ship's undoing.

Forced to focus on looks ahead of function, Henrik's team didn't have time to run stability tests on the ship. Instead, Henrik asked 30 marines to run from one end of the ship to the other. Even this simple test showed stability issues existed. But with Gustav demanding the vessel's delivery, Henrik could do nothing more than what he'd always done – comply with the order.

On August 10, 1628, The Vasa left the port of Stockholm carrying 150 passengers. Crowds of people, including several important dignitaries, gathered to send the ship off on its maiden voyage. The ship opened its gun bays to salute the crowd and prepared to fire...

That's when the wind hit.

A massive gust of wind caught the sails, tilting the ship to one side. At the same time, The Vasa fired its cannons into the water. The open gun ports let water in, and the ship began sinking fast. In just 15 minutes, The Vasa was underwater.

The tragedy killed 53 crew members.

And The Vasa only made it 120 metres away from the shore. Now, we want to ask you a question.

Do you think The Vasa would have sunk if Gustav didn't make constant changes to the project while forcing the ship builders to work faster?

We're willing to bet it wouldn't.

The story of The Vasa is a classic example of what can happen if you allow scope creep to affect your project. And while your digital product likely won't carry the same life and death consequences of Henrik Herbertson's ill-fatedship, there's still plenty that you can learn from it when it comes to developing your product.

Learning is exactly what we're going to do in this chapter. Here, we'll look at the critical elements of the Develop stage of the DevReady Process so you don't have to worry about scope creep scuttling your digital ship.

Element No. 1 - Create a Project Timeline

Project timelines are crucial.

Henrik Herbertson would tell you that. In his case, scope creep led to the inability to stick to the original project timeline. The Vasa was delivered six months later and over budget because King Gustav II didn't stick to the timeline created for development.

You can't allow that to happen with your solution.

Creating a project timeline allows you to figure out what you need to do, when you need to do it, and how long it will take. The timeline also serves as your project's development structure, allowing your tech team to focus on getting its work done rather than worrying about what it should be doing.

So, how do you create one?

Here's a quick breakdown of the steps:

1. Create a project scope statement that defines precisely what you're developing. Don't pull a King Gustav and start diverging from this statement once development begins. The DevReady process lets you nail down exactly what your MVP needs to contain, meaning any new ideas get reserved for future iterations.

2. Design a Work Balance Structure (WBS) that breaks down the time needed for each portion of the project. You'll likely break each milestone down into smaller chunks at this stage.

3. Break each work package detailed in the WBS down into tasks. Your tech team uses the set asks to guide them on what they need to do and when.

4. Determine any project dependencies. In other words, highlight the points where development can't continue until X, Y, and Z is completed.

5. Figure out how much time you need for each task. Your project engineer's experience should come in handy here as they'll hopefully be able to create some estimates based on projects they've previously worked on.

6. Identify the resources available for every task. By predicting resource availability early on, you get a more accurate timeline. If you need to commit your entire team to a specific portion of the project, you must accept that they can't move on any thing else until they complete that portion.

7. Highlight the critical milestones in the project. Use these milestones to track how far ahead or behind you are in the development work.

8. Compile all the above into a complete project timeline that takes your team from beginning to end.

Having a project timeline doesn't guarantee that you'll finish your project precisely where the timeline ends. Un anticipated issues may arise that divert resources, causing delays later in the timeline. But by having a solid timeline in place, your team knows what they're aiming for and can feel confident that you're not going to introduce any new requirements.

Element No. 2 - Make Sure Your Project Doesn't Contain Spaghetti Code

Do you remember Mellonie Francis?

Previously, we talked about how Mellonie built her tech team. She talked about trust being a vital component of making her team work.

But building a team isn't all Mellonie spoke about during her appearance on the DevReady Podcast. The Fine Book Beauty founder also talked about the problem of spaghetti code. She told us that early versions of the code behind her product had a lot of copied and pasted code, resulting in a solution that was so buggy she couldn't introduce it to the market.

So, what is spaghetti code?

Spaghetti code refers to any type of code that could cause your solution to fail. It's basically a complicated mess that results in reams of useless or inefficient code, creating the possibility of bugs and issues. This type of code usually results from a developer not understanding the language's specific programming constructs.

And there's another problem with spaghetti code...

It's a nightmare to revisit.

If you're planning on iterating on your solution, which is usually the case for a digital product, having spaghetti code makes the work harder. Your team can't jump straight in and make tweaks or additions. They need to spend an eternity trying to decipher code that barely works before they can do anything else.

How does spaghetti code occur?

Often, it's a case of developers jumping straight into their work without understanding the project scope or the frameworks their programming platform employs. It also occurs when inexperienced programmers use less refined programming structures to achieve results that would take a more experienced developer fewer lines of code to create.

If you think this sounds like a problem for your project engineer to confront, you're on the money.

Project engineers have the experience to ensure your solution doesn't contain complicated and broken spaghetti code. Combine that with the project planning you've already done, along with unit testing, and you ensure that this destructive code doesn't end up being the foundation that props your product up.

Element No. 3 – Create Automated Testing Procedures

We just mentioned unit testing.

Unit tests are one of the many types of small tests that you must build into the development process. Without these tests, your developers won't know if their code does what it's supposed to until they're thousands of lines in.

There's just one problem with testing:

Doing it manually adds a ton of time to your project.

That means you need to automate as much of the project testing as possible.

Start with the most repetitive tests. For example, the tests you run on new project builds don't vary much. You can automate many of those tests so your developers are free to focus on other issues. Any tests requiring several data sets or that are prone to human error are also good candidates for automation. If a test requires a lot of manual time and effort to complete, it's a good candidate for automation.

Next, think about the testing platforms you're going to use.

Let's say you're developing a mobile application for use on iOS and Android devices. That means you will need testing platforms that account for the differences betweenthe two operating systems. Plus, you may need to test to ensure your solution runs on older versions of each operating system so users who haven't updated aren't left out.

Finally, quality data and user interface change resistance are vital for automated tests. The data aspect is self-explanatory, with bad data leading to erroneous results that fail to show you anything about the product. Building change resistance into the testing scripts is a little more complicated. However, it allows your developers to continue tweaking the project's front-end while the automated tests highlight coding issues for them to confront.

Happily, many automated testing platforms are available for you to experiment with. Consult with your project engineer and leverage their experience to select a platform and create the scripts needed to test.

Element No. 4 – Make Sure Your MVP Fits the Bill

We've already discussed what a minimum viable product (MVP) is. Most of the early stages of the DevReady process help you to figure out what your MVP looks like. If you need a reminder, your MVP is the version of your solution that contains the minimum number of features necessary to impact the market. Once released, the MVP serves as the base for all the iterations you make over time.

With a defined MVP, you ensure that your product does what it says it will do. Having an MVP also enables you to get to market quick erand provides you with room to evolve the product based on customer feedback.

You've already defined your MVP if you've followed the DevReady Process.

So, what's it doing in the Develop stage?

The development process can reveal issues that you didn't anticipate in the earlier stages of DevReady. Some of these issues can affect your MVP, altering the project timeline in the process. To account for these issues, you must revisit your MVP repeatedly during development while asking a few questions.

First, does the MVP provide enough value for people to buy or use it?

It may have done so when you first defined your MVP. However, competing products and changes in the marketplace may shift the goal posts regarding what customers expect. If you can answer "yes" to this question, you're likely safe to continue development.

Second, does the MVP offer enough potential to retain early adopters?

Remember that an MVP is not the final product. In the modern tech world, it's rare for any product to reach a final stage. Your early adopters need to be able to see the potential for future improvements in your product, especially if you operate a subscription model. A lack of future benefits isn't necessarily a project ender. It just means that your MVP is actually the best possible product that you're capable of releasing.

And finally, does the MVP provide a feedback loop that can guide future development work?

An MVP's first purpose is to help you get something sellable to the market as quickly as possible. Its second, and often most underappreciated, purpose is to act as a vehicle for feedback from your early adopters. These initial customers will tell you what they want to see from the product in the future, which gives you the foundations for a roadmap that guides you through future product iterations. Your development process needs to provide the people who will use your product with a means to communicate their feedback to you, thus allowing you to iterate in a way that retains early adopters while increasing the product's appeal.

Element No. 5 – Focus on Future Scalability

A focus on the future is essential for an iteratively developed project. Your MVP is just the starting point. You also need to lay the foundations for future growth as part of the initial development process.

How these foundations look will differ depending on the specific project.

For some, a microservices approach to scalability will work. This often involves offering a sizeable primary service, which can splinter off into various sub-services that users can purchase as they need them. This approach allows your platform to handle more transactions while distributing activity evenly, so that no single element of the service becomes overburdened.

The one element of future scalability that appliesto all projects is ensuring availability.

The platform must operate with in an architecture that has distributed points of failure. In other words, the ideal product won't fall apart entirely if one aspect of it fails. Users will still be able to access different parts of the product, allowing you to maintain some availability while your tech team works on fixing issues. This type of architecture is also essential if you plan on adding new services as you scale your solution.

Poor Development Leads to a Poor Product

If you don't have a defined development process, your product will either fail or be so poorly made that the market doesn't embrace it.

The story of The Vasa demonstrates this.

Scope creep led to constantly changing development requirements, forcing Henrik Herbertson into a position where he had to make something that was beyond his capabilities. The result was a tragedy that humiliated the master shipbuilder and cost dozens of lives.

A good development process starts with a defined project timeline.

This timeline should not require much chopping and changing as your tech team works. If it does, you may need to go back a few stages in the DevReady process because you're likely not ready for development yet.

Good development is also efficient and based on the best practices in coding and testing. Finally, your team must maintain a future focus at all times. After all, you're working on an MVP here. If you're not laying the groundwork for expansionat this stage, that MVP will be all you can offer for a very long time.

By following the tips in this chapter, you create a smooth and effective development process that should result in a working product.

But you're not done there.

The unit testing you do through out development is not enough to ensure that your product is free of bugs and other issues. To ensure that, you need to implement Stage Nine of the DevReady Process – Quality Assurance.

CHAPTER

12

Stage Nine – Quality Assurance

The year was 1999, and all eyes on Earth were on Mars.

It was all systems go for NASA's Mars Climate Orbiter, a satellite set to be the first weather observer or biting a different planet.

But as the $125 million satellite approached Mars on that fateful day of September 23, it vanished—never to be seen again.

As it turns out, the satellite burned up in Mars' atmosphere before it could even land and begin its mission!

Needless to say, NASA's engineers were dumbfounded.

What there view board discovered next turned the situation from astonishing to downright ridiculous.

Apparently, the problem that cost NASA millions of dollars was an elementary conversion mistake. The engineers failed to convert their units from Imperial to metric!
See, the orbiter's thrusters were controlled remotely through multiple softwares. One software used the Imperial unit of measurement and calculated the force needed to be exerted by the thrusters in *pounds.*

Meanwhile, a different software that received that data and transmitted it to the satellite was setup with the metric unit of measurement. So, while it got the accurate figure calculated by the first software, it read the said number in *newtons*.

And since there are almost 4.5 newtons in one pound, the calculation of force was way off!

As a result, the orbiter went 105 miles lower into Mars' atmosphere than it was supposed to.

No wonder it burned right up!

The review board also found out that it was standard practice to convert everything to metric for space missions. It was so basic that an entire team of engineers assumed somebody else had done it, when no one did.

Just one Quality Assurance test.

Just one–and NASA could have saved itself millions of dollars and a whole lot of public ridicule.

Richard Cook, then the project manager for all Mars exploration projects said NASA made "big-time" changes after the incident.

Some planned missions were scrapped.

Engineers trooped back to the drawing board.

The agency's Mars program was rebuilt using strategies and concepts that had already been tested.

And never again did NASA experience the same epic failure in its Mars exploration efforts.

In fact, twin rovers Spirit and Opportunity were so successfully deployed that they surpassed every conceivable expectation.
They lasted 50 times longer than their original 90-day missions.

They delivered groundbreaking science discoveries.

And they inspired an entire generation of space engineers to always do their due diligence in order to see their best efforts shine through.

After you've developed your product, you'd be faced with a choice:

You could launch your software as fast as you can, or you could take the time to do Quality Assurance (QA tests) to make sure everything runs as intended.

When that time comes, we hope you remember the story of NASA's Climate Orbiter.

We hope you remember how it launched right to its death, when a single test could have saved it.

We hope you remember Spirit and Opportunity and how due diligence allowed them to smash all records set for their course.

And we hope you remember that you don't always have to learn things the hard way.

Certainly, we sure wish we didn't.

Our Deleted Database

Here's the story of how failing to do a QA test cost us an entire database.

This was how we used to set up our systems while developing virtual solutions several years ago:

First, we'd have a development copy of the website. It would be running locally on our own computers.

Second, we'd have a testing or staging server. That's where we push changes that are up for client review. Naturally, the dummy database used for this server was not the same as the actual database used for the live site.

And third, we'd have the live database and the live server, which our customers actually push out to the public.

So, we typically develop program codes by running them through these three separate setups. The code is first built in the development copy of the site, then we run the code for our clients on the staging server. And once approved, we swap the new code with the existing code on the customer's live website.

Now, swapping out program codes on a live website actually causes some downtime which could translate to lost business on the part of our clients.

So, a few years back, while we were working on an e-commerce solution for a caravan manufacturer, I was thinking of a way to eliminate that downtime.

I came up with a system where instead of just pushing out the new code to the live website, we'd first copy the customer's live database to our staging server and run the new code with it there. I reckoned doing this would mean we could deal with the downtime on the staging server so it wouldn't be a problem with the live site.

But here's the problem.

We got too excited to try out this new process.

I was like: *Well, what could go wrong?*

As it turns out, a lot of things could have gone wrong.

In fact, a lot of things did!

Remember: we were supposed to copy the client's live database to our staging server. But what actually happened was that the live database got swapped with the dummy database we were using in the development stage!

That means the new code was pushed out with the dummy data, and that's what went live!

Meanwhile, the real database got deleted on our staging servers!

That's right. Imagine my panic when that happened.

There I was, staring at a live, public-facing website that had all the wrong data, trying to reach out to the client on a late afternoon outside working hours.

It was a nightmare.

Luckily, I was able to get a hold of the hosting company. They were able to somehow retrieve a daily backup of the website as it appeared the day before.

Ultimately, the live website got restored and was working again in just a few hours. But going from panic mode to thinking I had to be up all night manually recovering the lost database, all the way to getting everything fixed was so stressful that I vowed to never again strayfrom a process that has been tried and tested to deliver good, quality results.

And as you prepare to develop your own virtual solution, I urge you to make the same commitment, starting with building a reliable Quality Assurance Team for your product.

How to Test Your Product Properly

A whopping $2.41 trillion.

That's how much developers lost from software failures in 2024 alone. And most of those losses could have been prevented by a thorough QA process.

We don't want you to make any contributions to that number for the year you launch your product, so here are some of the besttesting practices you want your QA team to follow while testing your product:

Plan the Right Tests by Analysing Your Software's Requirements

Your QA team should be able to design an appropriate test plan after analysing your software's functional and non-functional requirements. This test plan should comprise the following:

- The testing strategy that will be implemented for your product
- The scope of testing that will be done
- The budget required for the testing project
- The tools needed to report, track, and fix bugs

Each test should also clearly define the expected result so that all testers know what good results look like for each test case.

Set Up the Right Testing Environment

An important component of test planning is setting up the right testing environment for your software.

Imagine the kind of environment your product would likely be in once it hits the market. Think about the specific hardware, network configuration, and operating system your would-be customers would have when using your software. Factor in other characteristics like system settings and other software your product would be coexisting with once it goes live. Then, make sure your QA team tests your product in an environment that closely resembles that ideal set up.

On the other hand, if you're developing a product that fits a wide range of user profiles, make sure your QA team comes up with a test plan that like wise covers a wide range of scenarios. Have them test your product for different devices and operating systems, for starters.

Test One Thing at a Time

Once you've approved your QA team's test plan, it's time to set the stage for the execution of the tests.

And one of the best tips we can give you at this stage is to have each test focus on one objective, may it be speed, security, user interface, or any other metric.

Any bugs that may be detected while testing these components should then be submitted to your tech team through a defect-tracking system.

Run Regression Tests and Retests

Want to know why many products still fail despite undergoing a QA process?

It's usually because of instances where testers find and submit bugs, and then the development team fixes those bugs, and then everyone calls it a day.

What you should do instead is run a regression test every time something new is introduced to your program code. You see, these code changes and bug fixes may affect existing functions and create more bugs.

That's why your QA team should always be pushed to run regression tests to make sure the new code is still compliant with previously passed tests.

As your QA team reports bugs to your developers, the latter should issue a release notification. This release notice will detail all fixed bugs and newly implemented features to your software.

Your QA team must then identify all functionalities that are affected by these changes. And before approving the product for launch, they should design modified test cases to cover the scope of the new build.

This process should be repeated as many times as necessary to fix all bugs, recurring issues, and note all limitations, all while ensuring that the code still works as intended.

That's why we say QA needs to be a continuous process from start to finish. Your QA team should be aware of the actual product, design, functions, and features to ensure that they can test it appropriately.

You can't get testers who don't know how the system works. Because then, all they could do is just check for validation errors on forms. They won't have the capacity to check if logic is flowing as it should within the system and that things are transitioning appropriately.

To that end, you need knowledgeable testers who can run the five different types of testing you need to get done:

The Five Tests You Should Run to Ensure Software Quality

1. ***Unit Tests***

   As the name suggests, unit testing covers the smallest testable parts of an application. These units are tested individually and independently for proper operation.

   Unit tests are typically written by developers themselves as they write the code. This type of testing is powerful in detecting early flaws in code, which may be more difficult to locate in later testing stages.

2. ***Automated Tests***

   Running your code through automated tests may be considered another first-line defense against errors. After all, automated tests are very good at catching issues arising from changes in one part of the code that ends up creating errors on a different part of it.

   Note that automated tests are not always 100% guaranteed to resolve errors. Depending on how your code is written, automated tests may only be able to test certain things and not expose errors.

   Still, it pays to have this first line of defense before moving on to the next level.

3. ***Integration Tests***

   This is where you run multiple unit tests that are normally done in isolation on a chain. Doing so can expose any errors in the interaction between API and database website or mobile app to API to database. The chaining of the code along these elements is the main focus of integration testing.

   Another way to ensure that the code runs and transitions well across different elements is to build them into your continuous integration/continuous deployment pipeline (CICD) so that tests are automatically run… as new features are merged into your code.

   What happens here is if the tests do not pass, then the new feature doesn't get merged in. Instead, they have to be bounced back to the development team for resolution. This also ensures best results from your automated testing.

4. ***Automated UI Testing***

Regression errors happen when your developers end up writing new codes that reintroduce old bugs. Most of these errors happen in the UI, not in the logic. So, they need automated UI tests, which are tools that can do screen recordings and allow your developer to program tests where in you step through the website, click on buttons, and explore content like a regular user would.

So, as they check out the various features of the site, the program will record those steps and tweak the script as they input different values. This way, they can ensure that everything works as expected.

Automated UI testing should generally be done towards the end of a project, particularly when the UI is more stable and is only going through weekly changes. This is the time where you have to keep maintaining your automated UI tests.

5. ***User Acceptance Testing***

Finally, we've got User Acceptance Testing (UAT), which is also known as deployment testing. This is the last line ofdefense against errors becauseit will be done personally by you.

After accepting the code, you'd have to do your own level of testing on it to ensure that all features are executed as you envisioned them to work. Here, you'd be focusing on testing the logic and the UI/UX more than anything else.

And only after the code passes your User Acceptance Testing can you push through with production.

Test Hard and Test Often

A final note about running QA Tests: let your QA team do it as often as possible.

I know this chapter on Quality Assurance comes right next to Development, but I don't want you to think of these two things separately.

Testing should support your development, and thus the two things should occur in parallel.

Meaning, you shouldn't postpone testing until the end of your development cycle. In fact, you want your QA team to be involved in the development of your product as early as possible.

This way, your QA team can run tests with every new code update, identifying issues early before those bugs pile up. This helps ensure that the product you're developing is built correctly and won't need too many iterations later on.

So, test hard and test often.

By doing so, you can avoid getting stuck in a long and winding testing process and can instead go straight to the most exciting part for any development: product adoption.

CHAPTER

13

Stage Ten – Product Adoption

In 1999, Cosmopolitan had the brilliant idea to come up with its own line of yogurt.

Can you imagine?

Even at the time, the magazine company was already internationally acclaimed for celebrity news, fashion tips, and insights about sex.

So, why the foray into dairy?

Well, apparently, a survey conducted at the time found results that 65% of Brits used food during lovemaking.

For some reason, Cosmo thought this was a good enough reason to start producing their own yogurt at a hefty price of $1.17 per cup.

Any guesses as to how long the Cosmo yogurt stayed in the market before it was pulled down from the shelves?

18 months.

Obviously, the yogurt market was already saturated with several dozen brands, and Cosmo's readers were happy enough just reading the magazine. To this day, no one can tell if the Cosmo yogurt was actually any good—perhaps no one outside their QA team actually tasted any?

Now, here's another product that failed massively as soon as it hit the market:

Mobile ESPN.

It was so bad that when Steve Jobs introduced himself to then-ESPN President George Bodenheimer in 2006, his first words were:

*"Your phone is the dumbest f*cking idea I have ever heard."*

Ouch.

Mobile ESPN was the company's ill-fated attempt at launching a mobile service that's focused on sports.

The idea was great, though. A wireless service that lets sports nuts receive sports updates, browse ESPN content, and even launch GameCast on the go?

Sign me up!

Except not a lot of people did sign up, and for good reason.

See, the privilege of carrying sports in your pockets cost over $400.

Again, the year was 2006. Could you imagine forking over 400 big ones at the time for a phone that had some sports content on it?

Neither did ESPN's target market.

As such, Mobile ESPN is considered the largest, most embarrassing commercial disaster for the otherwise successful company.

By the end of 2006, not even a year since product launch, ESPN pulled the plug on Mobile ESPN.

But hey, Mobile ESPN is still not the record-holder in fastest product flops. A more viable contender for that title would be Facebook Home.

Remember Facebook Home? It was basically the social networking site's attempt at becoming the home screen for your phone!

And it flopped so hard Facebook decided to disband the team of engineers that originally worked on Facebook Home.

Turns out, nobody wanted zero control over what appeared on their phone's home screen!

The reviews were unanimous: Facebook Home was fine for a Facebook addict. But for everyone who's not a fanatic of the app, it was just weird.

Weird and expensive.

So, less than a month after its release, the two-year subscription plan to Facebook Home dropped from $99 to $0.99.

Now, what do Cosmopolitan yogurt, Mobile ESPN, and Facebook Home have in common?

Well, for starters, an entire team of developers, testers, and executives all thought these were great ideas. People spent good money developing, launching, and marketing these products. And yet, all three failed to get adopted by their intended market.

This is a very real possibility for any product that's about to hit the market. And to avoid the same fate, you need to first understand why seemingly good products fail in the adoption stage.

5 Common Reasons Why Good Products Fail

Unfortunately, a good idea is not all it takes to have a successful product launch. Take heed of the following reasons as to why some good products fail so you can prevent your product from going down the same path.

#1 : The Product Doesn't Create a Need

Remember in the Ideation stage when we defined the problem you want your software to solve? Well, I hope you were perfectly honest when assessing whether or not there's an actual need for the product you're developing.

After all, one of the most common reasons "good" products fail is that people don't see any need to buy it!

Take, for example, Microsoft's Zune. It was the company's attempt at competing with the Apple iPod. The Zune promised to do everything the iPod could, and nothing more. And that was its biggest mistake.

The Zune was an alternative to getting the iPod, but at a time when almost everyone and their mothers already had iPods, it offered no reason to make the switch.

#2 : The Product Solves the Wrong Problem

In1990, Maxwell House launched something called Ready to Drink Coffee.

This was before all the just-add-water coffee cups were invented, so you could imagine the product had great promise.

The premise packed a punch, too: the goal was to create a new, convenient way to enjoy coffee instantly without having to brew a cup at home.

All you have to do is buy a Ready to Drink Coffee from the convenience store, bring it home or to the office, microwave it for a few seconds, and you've got your cuppa ready!

So, why did the product flop?

Well, apparently, the product was only good on paper. In reality, you couldn't simply microwave the coffee in its original packaging. You still had to pour the product into a mug before putting it in the microwave.

And that's no different than pouring yourself a fresh cup of coffee from the coffeemaker! You could imagine the taste was worlds apart, too.

In the end, Maxwell House's Ready to Drink Coffee failed to solve a problem for its customers. All it did was present an intricate, overcomplicated way to tackle a non-problem.

So, before falling head over heels with your product idea, make sure it actually provides a working solution for a problem faced by your customers. Otherwise, back to the drawing board!

#3: The Product Assumes Wrong Things About Its Customers

Did you know that PayPal was originally developed to be an online banking service provider? But when it hit the market, users almost exclusively used it to make quick online payments. Luckily, its founders were able to quickly pivot their updates towards making PayPal more of a payment platform!

See, you can't always predict customer behaviour. They might end up using your product in away that you didn't intend. And that's okay. You just have to roll with the punches and focus on developing the aspects of your product that people are finding value in.

And whatever happens, don't stick with any wrong assumptions you may have made about your customers while developing your software. The moment you notice that your customers are behaving differently around your product, be ready to pivot towards giving them the experience they want.

Just imagine what could have happened if PayPal didn't make that pivot and stuck with its original idea of being an online bank. It wouldn't have developed its digital payments platform and it would have died a painful death in the hands of new competitors like Venmo or Cashapp since these products are more purposefully developed for that function!

So, don't work off of assumptions you have about your customers. The moment your product is soft-launched, watch closely how customers interact with it.

This way, you can design your software's next iterations towards what customers want to use it for.

#4: The Price is Not Right

Remember the Apple Lisa?

Can't blame you if you don't.

This personal computer was invented back when Apple was still responsible for some of the worst product flops of all time.

Still, the Lisa wasn't all that bad.

In fact, it was designed as a high-end computer for businesses. Its graphical user interface was ahead of its time, thanks to Apple spending a whopping $50 million in development.

Turns out, this hefty investment became the cause for Lisa's failure. See, Apple had to compensate for its development budget by slapping a $10,000 price tag on the Lisa. That translates to about $25,000 today—and was simply too high even for its target audience of business owners.

Apple was lucky enough to sell100,000 units in two years before ultimately discontinuing the Lisa in 1985.

So, remember that all hope for adoption is lost if your customers can't afford your product. People may be willing to pay a premium for superior products, but make sure you don't go beyond a reasonable price band for what your product offers.

#5 : The Idea is Poorly Executed

The reason we dedicated an entire chapter to Quality Assurance is because great ideas have been known to fail due to poor execution.

Things like bad design, poor user experience, and even a laggy interface can make or break the adoption of your product.

Remember why the 2007 Windows Vista is considered a failure? It used up so much power that most Microsoft users found it unusable. It made browsing such a terrible experience with slow-loading pages and missing graphics, it came as no surprise that just four months after the OS launched, new Dell computers were sold with an older version of Windows pre-installed.

The kicker is that a thorough QA test by a competent QA team could have prevented all that.

I know it can be a hard pill to swallow that the product you'd have spent time and money developing may fail once it hits the market. But being aware of that possibility just gives you the opportunity to iterate, redevelop, and try again.

Now, if you've checked your product against the five common reasons they might fail as enumerated above, you'd have a better chance at successful adoption.

Even so, there are still a few things you can do to improve your chances.

Improving The Chances of Adoption - The Key Tips

Here are three key tips that can increase the chances of your market adopting your product once it launches:

Key Tip #1: Run BetaTesting

Beta testing is one of the most beneficial phases of your software development. It is where you make your software available for a limited number of users before launching it to the general public. It serves a myriad of purposes, the most crucial of which is to gather feedback from a target group of users.

From there, you and your tech team can introduce improvements to your product before it is made widely available. Your would-be customers'point of view is the most important one to consider for your product to be considered a success, so if you want to improve your software's adoption rate, take the time to run a beta test.

Some of the things to watch out for during the beta test include:

- Whether or not all features of your software are functioning as intended
- Whether or not customers find your product user-friendly
- Whether or not there are bugs that your QA team failed to catch
- How your software performs in the real world
- The cost of deploying your software to the market

Overall, you should take advantage of the beta test to evaluate your software's performance in real-life conditions. While it is near impossible to develop a product that meets all its individual users' preferences, you should strive to find balanced, optimal solutions that satisfy the majority of your target audience.

As such, you need to provide a strong feedback platform for users who will be able to access your product during the beta test. Encourage them to share candid thoughts about their experience in using your product, and make sure you take all useful comments to your tech team for implementation.

One other way beta testing can help raise your adoption rate is through word of mouth. If early users have a satisfactory experience with your product, best believe they'd spread the word among their friends. This could help build up hype for your software even before its public release.

Key Tip #2: Understand the Customer Lifecycle

Adoption is all about consumer appeal. That's why you should commit to creating good customer relationships in order to build a loyal following for your product.

The first step to doing that is understanding the customer lifecycle. That refers to the various stages a consumer goes through before, during, and after completing a transaction with you. In other words, it's the journey your customer takes from where they are now—to a time in the future where they've purchased your product from you.

Before launching your product, you should be able to design a seamless strategy to nudge customers through each of these levels until they become an active advocate for your brand.

Here are the five basic stages in your customer's lifecycle:

1. **Reach**

   The first step in your customer lifecycle is reach. This happens after you've made your content and marketing material available where customers can find them. At this level, consumers develop an awareness of your product.

2. **Acquire**

   If you are successful in offering relevant content or messaging that resonates with your audience, you could easily nudge them in to acquiring the product you're offering. So, start with fully understanding your brand, the products you have on offer, and what type of person would be interested in buying them. This will help you craft personalised messaging that will, in turn, boost your conversion rate.

3. **Nurture**

   Keep in mind that just because you've got customers acquiring your software, it doesn't mean your job is done. If you want high adoption for your products, you should ensure that buyers are fully satisfied with their initial transaction with you. As with your beta users, make sure your customers have easy access to a feedback loop with you, so they can provide feedback that can help you improve their experience with your products.

4. **Retain**

   Higher adoption rests upon strong client retention. If you are able to develop a good long-term relationship with your buyers, the higher the chances are that they would return to you to make more purchases. To reach this point in the customer lifecycle, you need to be consistent with sending relevant and meaningful messaging to your customers. Keep them satisfied while using your software, and make them feel cared for with every interaction they have with you.

5. **Advocate**

Once a customer has reached there tention stage, the next step is for them to become an advocate for your product. This happens when a consumer is so satisfied with their experience, that they willingly recommend your software to friends and family.

If you are successful at building a loyal following, your brand advocates will help raise your product adoption rate by their own will.

That's the power of knowing how to take your users through the five-step customer lifecycle.

Key Tip #3 : Offer High - Quality Support

No product is perfect.

The earlier you accept that, the better you can support your early customers when using your software.

After all, we've all been in a situation where we needed help fixing an issue with a product or software we're using and we just can't get through customer service.

Frustrating, isn't it?

Personally, I've changed service providers for the sole reason that they've kept me waiting on the phone for more than a few minutes, or that they never get back to me when they said they would.

I'm sure you've done this, too.

So, when it comes to your own product, make sure not to lose customers due to poor tech support.

Make sure your team is available to answer questions and address issues via what ever channel your users might be most comfortable reaching out to you.

This includes providing tech support through:

- Phone
- Email
- Social Media
- Real-time chat

Being available to assist your users whenever they need it is something you will be appreciated for. Offering high-quality support gives you a golden chance to make a real difference with your customers' experience with your product. This way, you get to keep their loyalty for a long time, in turn keeping your product adoption rate at a higher level.

Raise The Adoption Rate

In reality, you can never predict how well your product will be received when it hits the market.

The most you can do is to execute your product idea as best as you can and hedge against some of the most common reasons why 'good' products fail. Make sure your product solves a real problem for an existing market, and make sure it is priced competitively.

After that, take a deep breath because your product will now be ready to see the light of day. At this point, you'll find out whether your product is ready for mass adoption.

Now, don't worry—if you've followed all the stages we've discussed so far to a tee, there's a chance that your product *will* be adopted.

Still, you could first run a beta test to see how well your product performs in the real world, gather feedback from early users, and make final tweaks before a final public release. From there, provide reliable support to your users as they go through the customer life cycle until they become advocates for your product.

And once you've got that loyal band of customers supporting you back and increasing your adoption rate with you, you'd be ready for the final stage of the DevReady Process – Commercialisation.

CHAPTER

14

Stage Eleven – Commercialisation

Would you rather slap Rihanna or punch Chris Brown?

That was the question featured in a Snapchat ad that went live in 2018. Snapchat must have thought they were being cheeky.

But obviously, their users thought differently.

See, not many people have forgotten the fact that Brown was literally arrested for assaulting Rihanna, resulting in one of the most high-profile domestic violence cases in recent memory.

The cheap shot was soap parent that Rihanna herself couldn't help but join the chorus of outraged voices that spoke out against Snapchat.

In a statement, the singer said:

"I'd love to call it ignorance, but I know you ain't that dumb! You spent money to animate something that would intentionally bring shame to DV victims and made a joke of it!!!"

Following the release of this statement, Snapchat's parent company lost a billion dollars in value as shares dropped almost 5%overnight.

It seems the marketing team that handled Snap's 2018 campaign didn't learn from the marketing scandal that happened a year before, as committed by Adidas.

Congrats, you survived the Boston Marathon!

That's the subject line of an email received by runners who participate din the 2017 Boston Marathon.

At any other time in history, it would have been a fine email subject. Those who ran the marathon would have appreciated the recognition, but the lesson to be learned here is that context always, always matters.

See, the 2017 Boston Marathon came on the heels of the 2013 Boston Marathon bombing, which killed three people and injured 250 more.

In other words, it seemed as if Adidas were congratulating those who weren't killed or injured in the bombing. As such, many people were offended. The shoe company eventually issued an apology for the poorly worded email, but the damage had been done.

As they say, hindsight is a wonderful thing.

Still, there were other marketing mishaps that frankly didn't need the benefit of hindsight. These are ads that make you go: *"Are we sure nobody in that creative meeting thought this was a bad idea?"*

Of course, I'm talking about Kendall Jenner's 2017 Pepsi ad.

Ah, to be the face of the most spectacular marketing fail of the year.

So, this is what happened. Apparently, Pepsi wanted to position its soft drink as a culturally unifying force of sorts. It was an ambitious goal, but there would have been a hundred different ways they could have achieved it.

Instead, *and we could only surmise that this is what happened,* someone at that creative meeting raised their hand and said:

"Oh, I know! Let's get Kendall Jenner on-screen to settle a stand off between police and Black Lives Matter protesters by giving a police officer a can of Pepsi."

Even more incredibly, this idea was given the thumbs up and the ad went live exactly as envisioned.

The result?

Ridicule.

Outrage.

And then-PepsiCo President Brad Jakeman had to step down about six months later, calling the Kendall Jenner ad "the most gut-wrenching experience" of his career.

This is why it bears repeating that it takes more than having a great product to actually be successful in converting sales. You may have an amazing software, but if you don't market it well, two things can happen:

It will either get lost in the shuffle, or worse, it will end up in a negative light the same way the brands mentioned above suffered from negative perception, albeit temporarily.

So, if you don't want your ads, emails, or even the mere mention of your software to leave a bad taste in your customers' mouths, you need to come up with solid marketing ideas to commercialise your product.

And that starts with nailing the basics of a sound marketing plan.

Here, we will discuss what a good marketing plan looks like and how coming up with one can save you the embarrassment and negative impact of a marketing campaign failure.

Nail the Five Ps of Marketing Plans

Also known as the marketing mix, the five Ps of Marketing are composed of variables you can control to satisfy your target market, add value to your software, and help you differentiate your business from competitors.

If you can make good decisions regarding all these variables, then you're already five steps closer to marketing your software right:

1. **Product**

   Simply put, product refers to the software you're set to develop. Product-related marketing decisions you need to nail include:

   - UI/UX
   - Branding and Positioning
   - Support
   - Security

   If you have a well-executed product idea, good marketing collateral includes branding and positioning, UX, UI, website, warranty, packages, and support packages. Additionally, you want your marketing to highlight the features and benefits your customers can get from buying your software.

2. **Price**

   Remember the Apple Lisa? It was a product that could have very well succeeded if it were a few thousand dollars cheaper.

   This tells us that your pricing decisions can make or break your software's chances for success. But your product's price tag is not the only price-related decision you need to seriously consider. You've got to think about discount options, payment arrangements, and even credit terms that could help you sell your software more easily. For instance, if it's a monthly product, you need to consider offering discounts for annual subscriptions. And if you're offering it to either one or more users, you can either offer a discount or increase the price, depending on the features you open up to them. More features, more users. For example, a collaboration feature doesn't need to exist for a solo user. But for 10 users, it adds value, which means a higher price point.

Note that a lot of market research is required before you can decide on a pricing strategy. You need to understand where your product stands in the current market, and how your competitors are pricing similar products.

We'll talk more about developing your pricing model later on in the chapter, but for now, let's move on to the third P.

3. **Promotion**

The meat of marketing is promoting your software through activities that make it more known to consumers. Sponsorships, ads, and partnerships are some of the things you need to plan in order to raise awareness for your software.

Remember when we talked about the customer lifecycle? Promotion is what will help you move customers through each of those stages until they become active advocates for your product and brand.

Since paying for ads can be costly, you'd want to carefully analyse the effectiveness of running them in relation to how much sales you can generate from each ad campaign.

4. **Place**

Place-related decisions include distribution channel sand how they can help you get your software to your target customers.

You have to make your product accessible to customers so they can easily find it and purchase it. Since you're offering software, you have several unique distribution options to consider. For example, you might choose to distribute your product through online platforms suchas your website, app stores (like Google Play or the Apple App Store), or third-party marketplaces (such as Amazon or specialised platforms tailored to your target industry). These channels allow you to reach a global audience with minimal overhead costs.

Alternatively, if your software is designed for businesses or enterprise clients, you might leverage partnerships with value-added resellers (VARs), managed service providers (MSPs), or even direct sales teams to ensure your product reaches decision-makers. Another key consideration is whether your software will be available as a downloadable product, a subscription-based service hosted in the cloud, or a combination of both.

At any rate, you want your products to be available to customers at the right time, at the right place, and in the right quantity. This means ensuring that your digital distribution channels are reliable and scalable, so users can always access your product without delays. For instance, using a robust content delivery network(CDN) can help minimise latency and ensure fast downloads, while secure payment gateways and automated licence management systems streamline the purchasing process. By optimising these elements, you create a seamless and satisfying customer experience that encourages sales and builds trust in your brand.

5. **People**

Your staff, sales representatives, and tech support team are some of the most important people on your side at this point of the DevReady process. All client-facing staff members form part of your marketing, and they can either help or harm your marketing efforts.

That's why you need to make careful decisions when it comes to your staffing, especially with your customer service. Always think about how you want your brand to be perceived by customers, and make people-related decisions based on that.

Generally, you want friendly and reliable staff members who offer exceptional support and customer service to your users.

After nailing the five P's of your marketing plan, you're now ready to create a pitch deck for your product, so you can start selling it to customers and potential investors.

Create Your Pitch Deck

Imagine you're invited to a convention attended by hundreds of people, some of whom may be interested in investing in your idea. You're given just 15 minutes to present your product. In that time, you need to introduce your software, highlight its key benefits, and appeal to the needs and preferences of your audience—all while establishing rapport and aiming to secure their investment.

One tool that can help you deliver a compelling presentation is a well-crafted pitch deck.

A pitch deck is a focused presentation, typically 10 to 5 slides, that delivers the essential details about your product, business model, and vision. Importantly, your pitch deck should be tailored to your audience. For example:

- **An Investor Pitch Deck** focuses on the financial and market potential of your software. It addresses aspects such as the opportunity, businessmodel, competitive landscape, and use of funds.

- **A Sales Pitch Deck** caters to specific buyer personas. For instance, if you're presenting to a CTO, the deck might emphasise security and governance features. On the other hand, if you're targeting a Head of Innovation, your focus would shiftto howthe software drives efficiency and improves success rates for new initiatives.

Zooming in on creating a deck for investors, here's what your **Investor Pitch Deck** should include:

- A briefs ummary of who you are and your expertise.

- An overview of your software and the big idea behind it.

- Your target market and ananalysis of TAM (Total Addressable Market), SAM (Serviceable Available Market), and SOM (Serviceable Obtainable Market).

- The problem your software solves and why this problem is significant.

- The opportunity in the market and how your product capitalises on it.

- The key benefits and advantages your software offers.

- A clear explanation of your business model and how investors can achieve a return on their investment.
- Your use of funds — how the investment will be allocated to drive growth.
- A roadmap that outlines your short - and medium - term goals.
- Realistic financial forecasts, such as a 24 - month projection, with a focus on achievable outcomes.

While creating your deck, aim to keep it concise and engaging. Avoid overwhelming your audience with excessive detail or lengthy slides. Remember, no one has time to go through a 50-slide presentation that feels more like an information dump than a persuasive pitch.

Each slide in your pitch deck should contribute to answering key questions, such as:

- What's the problem you're trying to solve?
- How big is this problem, and what's the market potential?
- What's your proposed solution, and how is it unique?
- What benefits will users gain, or how will investors see returns?
- How do you plan to use the funds?
- Who is your team, and do they have the skills and experience to deliver?

If a slide doesn't contribute to answering the sequestions, consider removing it or replacing it with more impactful content.

Finally, remember that your pitch deck isn't just about data and facts—it's about telling a story. Investors are more likely to support a product that resonates emotionally and aligns with their values. Use your presentation to create a narrative that demonstrates your passion, vision, and the transformative impact your software can have on the market.

With a targeted and thoughtfully designed pitch deck, you'll have the foundation you need to confidently approach investors and inspire them to believe in your vision.

Forming Strategic Partnerships

Have you ever had a cup of Starbucks inside a Barnes & Nobles bookstore? Or streamed your Spotify playlist while riding an Uber?

Perhaps, you've seen someone sporting a Louis Vuitton bag that perfectly matches their BMW i8?

If you've done any of these things, then you already have a pretty good idea of how strategic partnerships work. It's when two non-competing brands that have related goals and a shared audience band together for products or programs that help boost their respective marketing and adoption efforts.

I mean, good coffee anda good read always go hand in hand. A car ride in bad traffic can be made more bearable with the music you like. And those who can afford a BMW sports car can probably also afford an LV bag.

That's why all those strategic partnerships we've mentioned earlier worked so well – they provide a strategic solution that markets both brands in a positive light.

Likewise, you want to engage in partnerships that can boost your marketing and promotional efforts. For instance, you could think about having a Partner Program. It's where you target innovation consultants who can open up doors for you. You partner with them in promoting your product that supports innovation in a company. These partners already have established relationships, which means they can help increase your chances of conversion.

For this, you need to identify the right companies to have strategic partnerships with. Remember that your goal is to position your software in a positive light, so avoid inking agreements with companies that have less-than-ideal reputations that might also affect yours. Also, identify companies that serve the same market as your intended audience—the same set of people you want to get in front of.

And remember, there are costs involved in diving into these partnerships. You don't want to waste your hard-earned money partnering up with companies that won't get you closer to your target audience. You also have to carefully evaluate the benefits you stand to gain and the risks you might face by entering into such agreements.

Another tip I can give you is to starts couring for potential partners earlyon in your development process. You don't have to wait until the product is ready for launch before planning these partnerships. After all, you might find that some alliances can help you raise awareness and generate interest in the product you have in development before it even hits the market.

Remember what we said in the previous chapter: there's no such thing as 'too much' when it comes to efforts that will accelerate mass adoption of your product. If you can find strategic partners that can boost your software's adoption rate without breaking the bank, go for it.

Developing Your Pricing Model

Now, you're not the only one who doesn't want to break the bank to get your product out there.

Your target customers would want to avoid it, too.

That's why you need to develop a pricing model that works for both you and the market. Remember that no amount of good marketing can make up for an overpriced product whose price tag cannot be justified.

To be frank, pricing a new product is one of the most complicated marketing decisions you have to make. You should start working on your pricing model long before you plan to launch your product on the market.

Earlier, we talked about beta testing and how it can help you fine-tune your product before its general public release. Well, beta testing also gives you an opportune time to come up with a reasonable pricing model for your software.

Here's what you can do:

Step #1: Identify a Sample

You'd want your beta testing sample to be composed of knowledgeable users who can give you useful feedback with regard to your product and the price you want to offer it for. These are people who would likely buy your product as early adopters. They are people who will put your software through rigorous use and thus stand to benefit from it the most.

Step #2: Educate Them About Your Prototype

While you have leading-edge thinkers and innovators as part of your sample, you still have to educate them about your software.
Why?

Because you need to gain an insight into the cost of educating customers on how to use your product. If the cost of producing manuals, instruction kits, or demo kits would push your price point to unreasonable levels, you might need to re-engineer your software.

At any rate, you need to know the costs of educating your customers because that will factor into your pricing model.

Step #3: Watch Out for Value Gaps

During beta testing, make sure to ask your customers to identify improvements they'd like to see with your product. If feasible, have them rate the individual attributes of your software to find out how satisfied they are with your development.

Having this information not only helps you plan a reasonable pricing model for your product, but also helps you map out future iterations or patches for your software.

Step #4: Measure the Value Delivered by Your Software

Once you've gathered enough information about your sample's experience using your software, then you can measure the value your software delivers to your users and build a price around that.

Simply put: the price your customers have to pay for your software should be commensurate to the value or benefit they get from it.

Step #5: Develop Your Pricing Model

Finally, you should develop your pricing model through a rigorous elimination process. First, rule out unfeasible ranges and come up with a spectrum that is reasonable enough that people would be willing to pay for it, while also allowing you to make money from your development.

If you find that the full costs of engineering, testing, launching, and marketing your product are too astronomical and that they require you to price your software beyond a reasonable range, find ways to slash your operating costs.

Once you have a feasible price range that lets you cover your operating costs and desired profits, then all that's left to do is come up with pricing packages and promotional offers you may use to gain your first-ever buyers.

Go to Market in the Right Way

Going commercial with the software you painstakingly engineered, developed, and tested is truly one of the most exciting and rewarding parts of your software development process. However, you'll find that it's also the most nerve-wracking process you have to go through.

After all, the public is always a harsh critic. What sounds good within your marketing and creative meetings may actually be deemed offensive, ridiculous, or downright laughable in the real world.

Remember Kendall Jenner and her Pepsi ad? Or Snapchat's 'Would You Rather' ad? And the fateful 'Boston Marathon Survivor' email?

All are examples of well-meaning marketing work that proved disastrous. Now, as you work on your marketing plans, pitch deck, strategic partnerships and all other ways you plan on marketing your software, just always think of how you want your users to perceive the software you invested blood, sweat, tears, and money on.

Market your software the right way, and you could complete the software development process with flying colours. Doing this right is the only way you can reach your target market, strike the right chords with your intended audience, and launch your tech solution successfully — all the way to mass adoption.

CHAPTER

15

Get DevReady!

Before we conclude, let's look at two examples of how the DevReady Process has helped businesses bring their ideas to life and achieve measurable results.

Russell O'Connell: Scaling with Confidence

Russell O'Connell and his team had been running a successful home services business for over 20 years, initially in the UK and more recently in Australia. As they prepared for a major expansion, they needed to digitise their well-established processes and find a way to scale their operations efficiently.

After interviewing several development companies, Russell chose Aerion Technologies for our clear and systematic approach. As Russell explains, "The chemistry was great, and they really took the time to understand our business. Their logical, step-by-step process resonated with how we run things ourselves."

Through the DevReady Consultation, Russell's team worked closely with Aerion to map out their application requirements. Using Figma prototypes, they were able to design user flows and features in real-time. "The live collaboration saved time and ensured we were all aligned," says Russell.

The partnership didn't end with planning. Aerion's deep understanding of Russell's business allowed them to offer valuable suggestions that enhanced the app's functionality. This approach kept the project on track and ahead of schedule.

For Russell, the process delivered exactly what his business needed: "We now have a clear architectural plan and a reliable development process. That's what every business needs to feel confident moving forward."

Surge Singh: Refining Without Complexity

Surge Singh, co-founder of Locumate, a marketplace connecting pharmacies and locums in Australia, already had an app in place. However, as his business grew, he needed to refine and scale it without adding unnecessary complexity.

After connecting with Aerion Technologies, Surge found the guidance he was looking for. "Their structured approach to development made everything feel clear and manageable," he says.

Aerion's product manager worked with Surge to prioritise features that aligned with user needs while avoiding overcomplication. "They helped us focus on what really mattered," Surge explains. "It allowed us to refine the app and introduce new features without compromising simplicity."

Aerion's thorough process—from translating technical concepts into simple terms to ensuring rigorous testing—helped Surge's app scale effectively. "For a non-technical founder, having a team that simplifies the process and ensures quality is invaluable," he adds.

The DevReady Advantage

Russell and Surge's experiences highlight the practical benefits of the DevReady Process. It provides a clear, methodical approach to development, ensuring every decision is intentional and aligned with your goals.

With DevReady.Ai, the process becomes even more streamlined. The platform enables you to generate development-ready plans quickly and with the same level of precision and clarity as our consultations. Whether you're launching a new idea or scaling an established business, DevReady.Ai accelerates your journey from concept to execution.

Let's Build the Future Together

Developing a solution is a challenging process, requiring technical expertise and strategic foresight. Even the biggest names in tech have learned hard lessons—Apple's Lisa computer and Google Glass are reminders that innovation must align with user needs and market readiness.

For startups and small businesses, trial and error can be costly. The DevReady Process is designed to provide a clear path forward, reduce risks, and maximise your chances of success.

We're so confident in this process that we offer a money-back guarantee: If you don't feel more confident and clear about your project after our consultation, we'll work with you until you do—or refund your money, no questions asked.

Wherever you are in your product journey—whether refining an existing app or starting from scratch—DevReady can help you move forward with confidence. By combining DevReady.Ai with personalised consultations, we provide the clarity, direction, and strategy needed to make your product market-ready.

Let's get started today. Together, we'll turn your vision into a fully developed solution that delivers results.

DevReady.Ai

Over the past few years, building software has become significantly more accessible. Tools now exist that allow people to generate interfaces, write code, and deploy applications in a fraction of the time it once took.

That shift has changed how quickly something can be built. But it hasn't changed what needs to be understood before it is built.

In many cases, the speed of these tools introduces a different kind of risk. It becomes easier to move forward without fully defining the system, without aligning on what is required, or without understanding how the pieces are meant to work together. The result is often something that functions but doesn't quite hold.

DevReady.Ai sits in front of that.

It is an AI-powered platform designed to help structure the thinking that comes before development. The platform guides you through defining your idea, shaping the system, and working through the details that are often left implicit or assumed.

This includes:

- clarifying the purpose of the product
- defining features in a way that can be built
- establishing how users interact with the system
- outlining the logic, constraints, and expectations that sit behind it

As you work through the platform, these elements are captured and refined into a coherent plan aka a DevReady Brief which is a set of modules, user stories, and tasks. It can be handed to a development team, or used alongside modern tools, with a clear understanding of what is being created and why.
The intent of DevReady.Ai is not to slow the process down. It is to ensure that what moves quickly is also moving in the right direction. When the underlying

thinking is clear, the tools used to build become far more effective. Without that clarity, speed tends to amplify whatever is missing. DevReady.ai provides a way to establish that clarity early, so that everything that follows has a stronger foundation.

Further information and access to the platform can be found at:

www.devready.ai
Or just scan the code below:

Working Together

By this point, the ideas in this book should feel familiar. Not as a set of steps to follow, but as a way of thinking about software more clearly.

For some teams, that shift in thinking is enough. It changes how decisions are made, how problems are approached, and how systems are shaped over time. In other cases, applying that thinking in practice is more difficult. There are competing priorities, existing constraints, or simply too much uncertainty around what the right next step should be.

That is where working together becomes useful.

At Aerion Technologies, the focus is on helping teams bring clarity to what they are building. This often begins before development, where the goal is to define the system properly and ensure that the direction is well understood.

From there, the work can take different forms. Sometimes, it involves shaping a new product from the ground up. Sometimes, it involves stepping into a project that has already started and needs to be redirected. In other cases, it involves rebuilding something that was never clearly defined in the first place.

The context changes, but the approach remains consistent. Clarity is established first. Decisions are made with intent. Development follows from that foundation.

When that process is in place, the outcome tends to be more predictable. The system is easier to understand, easier to build, and more likely to deliver what it was intended to.

If you are looking to get DevReady, and a conversation would be useful, you are welcome to get in touch.

www.aerion.com.au

info@aerion.com.au

1300 134 591

Or just scan this

The starting point is simply to understand what you are working on and what the right next step might be.

Made in the USA
Monee, IL
04 November 2026

67385950R10105